T0138374

Instrumental Biology
or The Disunity of Science

SCIENCE AND ITS CONCEPTUAL FOUNDATIONS
David L. Hull, editor

INSTRUMENTAL
BIOLOGY
OR
THE DISUNITY
OF
SCIENCE

Alexander Rosenberg

THE UNIVERSITY OF CHICAGO PRESS
Chicago and London

ALEXANDER ROSENBERG is professor of philosophy at the
University of California, Riverside. He was co-winner of
the 1993 Lakatos Award in philosophy of science for his book
*Economics—Mathematical Politics or Science of Diminishing
Returns?* also published by the University of Chicago Press.

The University of Chicago Press, Chicago 60637
The University of Chicago Press, Ltd., London
© 1994 by The University of Chicago
All rights reserved. Published 1994
Printed in the United States of America

03 02 01 00 99 98 97 96 95 94   1 2 3 4 5

ISBN: 0-226-72725-4   (cloth)
　　　0-226-72726-2   (paper)

Library of Congress Cataloging-in-Publication Data

Rosenberg, Alexander, 1946–
　　Instrumental biology, or, The disunity of science / Alexander
Rosenberg.
　　　　p.　　cm. — (Science and its conceptual foundations)
　　Includes bibliographical references (p.　　) and index.
　　1. Biology—Philosophy.　2. Science—Philosophy.
　　3. Instrumentalism (Philosophy) I. Title.　II. Title: Instrumental
biology.　III. Title: Disunity of science.　IV. Series.
QH331.R66　　1994
574'.01—dc20　　　　　　　　　　　　　　　　　　94-17083

⊗ The paper used in this publication meets the minimum
requirements of the American National Standard for Information Sciences—Permanence of Paper for Printed Library
Materials, ANSI Z39.48-1984.

For

EVERLY BORAH FLEISCHER

Contents

Contents

Acknowledgments

This book is the result of ten years of thinking about the philosophy of biology from the time I completed the manuscript of *The Structure of Biological Science*. I was in many ways content with that account of biology and especially the relation between molecular biology and the theory of natural selection. But in the years since its appearance, I have come to realize that there were important lacunae in my discussion and that some of my views did not really reflect a stable conceptual equilibrium. Moreover, these views were subject to interpretations I *did not* intend and criticism they *did* deserve. It was the combination of a certain molecular biologist's dissatisfaction with *The Structure of Biological Science* and two philosophers' interpretations of its theme that first goaded me into thinking systematically about how to fill these lacunae and what filling them would mean for my broader view about the nature of biology. The philosophers to whose interpretive efforts I owe my motivation are C. Kenneth Waters and Paul Thompson.

The result of their efforts is a work far more weighted toward the philosophy end of the spectrum of the philosophy of biology than was *The Structure of Biological Science*. If the difference between philosophy and science is merely one of degree, as philosophers since Quine have come to recognize, this book will still have considerable relevance for biologists. Nevertheless, disputes in the philosophy of biology increasingly ramify for problems about the methodology of other disciplines and, for that matter, problems about epistemology and metaphysics generally. This makes more pressing a work that seeks to reconcile the lessons learned about biology's differences from other empirical sciences with an empiricist epistemology and a physicalist metaphysics.

I have profited from discussion and detailed criticism of drafts of the entire manuscript by John Dupré, Kim Sterelny, and C. Kenneth Waters. I am also indebted to John Dupré for providing me with page

x
Acknowledgments

proofs of *The Disorder of Things: Metaphysical Foundations of the Disunity of Science* at the time I was revising this manuscript (but after I had given it its subtitle). The role of Dupré's work in what follows should be evident.

My greatest debts for what I understand about the *philosophy* of biology must accrue to several philosophers whom I criticize hereafter most persistently: John Beatty, Robert Brandon, John Dupré, David Hull, Harold Kincaid, Philip Kitcher, Elisabeth Lloyd, Michael Ruse, Elliott Sober, Kim Sterelny, Paul Thompson, and Kenneth Waters.

CHAPTER ONE

Biology as an Instrumental Science

IS SCIENCE AN enterprise that aims to limn the structure of nature and reveal the essence of things? Or is science ultimately a practical means of controlling our environment, physical, chemical, biological, or social? Philosophers debate these alternatives—science as an account of reality (realism) versus science as a useful tool for coping with it (instrumentalism)—but the working lives of scientists are rarely touched by this dispute. I argue here that the debate between these two approaches, realism versus instrumentalism, the theoretical versus the practical, has direct relevance for the ambitions of the biological and behavioral sciences. If I am right, biology and all the disciplines that rest upon it—psychology and the other human sciences—must aim to be practical rather than theoretical. If my analysis of biological method and theory is right, biologists and social scientists should find relief from the frustrating demand for general theory and recognition for the inevitably applied character of their disciplines. And philosophers will be able to reconcile their epistemology with the evident differences between biology and physics.

Biological science is an instrumental science to a much greater degree than the physical sciences (and to a lesser degree than the social and behavioral sciences). That is, it should be viewed much more as a useful instrument, a collection of heuristic *devices* and useful rules of thumb, than the physical sciences are. Its most well established theories should not be treated the way we treat well-confirmed theories in chemistry and physics—as a set of best guesses as to the truth about the way the world is, independent of us. It should be treated mainly as a collection of useful instruments for organizing *our* interactions with the biocosm. Viewing biology this way will enable us to improve our understanding of its structure, its epistemology, and its relation to the rest of science—physical, behavioral, and social.

My thesis is that biology is *relatively* more instrumental a science than the physical sciences. In order to explain it I need to explain what one might call *absolute* instrumentalism and distinguish my relative thesis from it.

INSTRUMENTALISM VERSUS REALISM

Nonrelative or absolute instrumentalism is the thesis that our scientific theories do not report the way the world is but are instruments for organizing our experience and predicting the consequences of events, states, or conditions to which our observational apparatus (our senses) is sensitive. Instrumentalism is traditionally embraced by philosophers and others committed to a strict construction of empiricist epistemology. According to the empiricist experience, data, phenomena, and observations fix the limits of knowledge. Taking instrumentalism seriously requires its proponents to explain or explain away the apparent commitment of scientific theories to unobservable entities, properties, states, events, processes, and conditions. One traditional approach to doing so, associated with logical empiricism, is to attempt to define theoretical terms by way of nontheoretical ones. For instance, density may be defined explicitly in terms of volume and mass, both quantities some of whose values we can determine by observational means. Given such definitions, we can translate sentences of a theory that advert to theoretical properties like density into ones that refer to and describe observable things and properties only. The trouble with this program for reconciling instrumentalism and scientific theory is that in most cases the required definitions cannot be provided. The theoretical terms of physics in particular—terms such as quark, charge, electromagnetic field—do not admit of adequate definition in terms of observables. Nor do chemical terms like acid, base, catalyst, ion, and so forth.

Nowadays, a more popular alternative defense of instrumentalism accepts that the scientific theories we believe really do commit us to theoretical entities, and therefore as consistent empiricists we should adopt an agnosticism about these claims, accepting as true only science's claims about what can be observed. Even though scientific theory requires an appeal to theoretical entities to "save the phenomena," we need to assess them for acceptance only with respect to their empirical adequacy—their adequacy as tools for systematizing our observations.[1]

1. One leading exponent of this view is Bas Van Fraassen, who describes his thesis as "constructive empiricism" (*The Scientific Image* [Oxford: Oxford University Press, 1979]).

One problem that faces both sorts of instrumentalism is the popular view that we cannot identify a body of terms that is purely observational and entirely neutral between theories, and that we can use to describe phenomena and so impartially test competing scientific theories. Without such an observation language, however, the limits of knowledge that empiricism stakes out become problematical. As any student of the philosophy, history, or sociology of science will testify, instrumentalist treatment of the claims of science has been combined by some with a denial that there are any unimpeachable observational findings. The combination has provided a basis on which to reject science's claims to be able to secure objective truths about a world independent of our conceptualizations and constructions. Thus it is but a short step from this sort of instrumentalism to the widely held claim that, since there are no theory-free descriptions of observation for scientific theory to systematize adequately, scientific adequacy is always and only relative to our interests and interpretations.

The contrasting thesis to instrumentalism is labelled scientific realism by its proponents. It is the thesis that the claims of theoretical science are either true or false, and that the most reasonable belief to embrace is that the most well confirmed of our scientific theories are not merely empirically adequate but true in their claims both about observations and about the underlying entities, states, conditions, and properties that the theories advert to, in order to explain and predict observational phenomena. When challenged to provide a justification for beliefs about theoretical entities and properties that cannot be observed, scientific realists respond in a variety of ways. Some hold that to believe the truth of the theory that best explains available observations is an argument form acceptable to empiricism. Other proponents of scientific realism eschew empiricism, and even normative epistemology, cleaving to science as the court of last resort with respect to what we should believe about the ultimate constituents of the universe.[2]

Realism, as I understand it, holds that our theories should aspire somehow to correspond to the world, and that the greater degree of correspondence the more adequate the theory: ultimately for each kind of thing, property, or relation that obtains in the world there ought to be in our theory a predicate—a natural kind—naming that kind of thing, property, or relation, and the relations of cooccurrence, succession, and mereological dependence between things in the world should

---

2. Somewhere W. V. O. Quine writes, "Philosophy of science is philosophy enough."

be mirrored in the most adequate theory. This theory may well be unattainable, but it is what science aspires to. Realists can and should be instrumentalists about some parts of science. They need not deny that science advances through the provision of models, idealizations, unrealistic simplifications, and other heuristic devices. Nor need a realist deny that scientific explanation is heavily constrained by contextual considerations that reflect human interests and human knowledge. But besides these "concessions" to science as a human instrument, the realist also insists on the truth of some scientific claims about the world independent of our beliefs, interests, or even our existence as cognitive agents. Between epistemology and metaphysics the realist gives priority to the latter: there are absolute truths about the way things in the world are disposed, regardless of the limitations on any sentient creature's abilities to discern these truths. In fact, we humans have been able increasingly to discern these truths as reflected in the growing *correspondence* between our theories and the world. Indeed, this is the best argument for realism. Without this history of successive approximation to the truth about the world reflected in our science, the fact of our amazing technological advances would be a transcending mystery, a cosmic accident.

The instrumentalist rejects all talk and thought of correspondence. At best, correspondence is beyond our powers to establish; at worst, it is a meaningless illusion. According to the instrumentalist, theories are shaped solely to meet our human needs, interests, and limits. The most adequate theory is the one that most fully meets these needs, interests, and limits. A theory does so when it can accurately predict how our observational apparatus—the five senses characteristic of *Homo sapiens*—will be successively affected. Since our knowledge does not extend beyond our senses, claims about objects too small or too large to be detected are either to be understood as heuristic devices for systematizing our experience, as open invitations to skepticism, or finally as meaningless nonsense. Any claim about the existence or character of such unobservables is metaphysics, in a pejorative sense according to which that term connotes idle speculation without empirical content.

My thesis about the cognitive status of biology does not require that we take sides on, let alone resolve, this dispute between instrumentalists and scientific realists.

Instrumentalism in One Science

What I argue is that the character and structure of biological theory more nearly reflect these human needs, interests, and limits than does physics or chemistry. Relatively speaking, biology is more instrumentalist a science than these latter disciplines are. Another way to put my thesis is this: Suppose you are an instrumentalist about scientific theories. Then independent of whatever purely epistemological reasons there may be for claiming that all scientific theories are but useful instruments, there are further *biological* reasons for supposing that biological theory is at most a heuristic device.

Alternatively, suppose you are a realist and believe that the theories of physics and chemistry are more than merely useful fictions, that they are approximately true of the world both in their theoretical claims and in their observable ones, and that we are capable of increasing their degree of approximation to truths about the world, which obtain independently of our sensory knowledge of it. Then my thesis is that there are still good reasons to conclude that biological theory is an instrument reflecting significant limitations on the conceptual and computational powers of human beings. Accordingly, the set of theories that maximize simplicity and strength in systematizing the sequence of events in the history of the universe will not include anything we would recognize as biological theory, except for a very abstract and generic version of the theory of natural selection.

Or again, biology is more of an instrumental science than physics and chemistry in this sense: if our cognitive and computational powers were vastly greater than in fact they are, biological theory would be much different from what it is, while physical and chemical theories would not be so different from what they are. If our computational and cognitive powers were much greater than they are, we would still need the second law of thermodynamics in order fully to explain and predict a wide variety of phenomena, but we would not need most of what we now consider the interesting generalizations of biology. There are interesting generalizations embodied in biological theory that we would miss if we eschewed the descriptive vocabulary of biology—its "natural kinds"—but these are generalizations in part about us and our epistemic resources, as well as generalizations about the world. For agents more cognitively and computationally powerful than we, these generalizations would be explanatorily otiose. As such, the generalizations and theories of biology are, in an old-fashioned way of putting it, "mind-dependent."

If there were cognitive agents much smarter than we are, who could keep far more alternatives clearly in mind and trace out the implications of far longer and more intricate chains of reasoning than we can, then the generalizations of biology would be very different from what they are, and no one would reckon what we take for the significant generalizations of biology to be any serious part of that science. The character of biology is contingent on our cognitive and computational powers; it reflects as much the "grain" of description of nature that we can handle as it does any "grain" to be found in nature. It is in this sense that biological theory is a heuristic device reflecting our needs, interests, and powers.

Still another way to express my claim is this: the universe is simple enough in its operations from the level of microphysics up to the level of the organic molecule that creatures of our intelligence have been able to discover the relatively simple general laws—the nomological generalizations governing these relatively simple processes. To do so we have had to avail ourselves of models, analogies, approximations, idealizations, and other heuristic devices. But in physics and chemistry these instruments have helped us uncover a small set of simple laws governing a vast range of phenomena.

For causes that I shall sketch in chapter 2, once matter aggregates beyond the level of organization of the biologically active macromolecule, the level of complexity becomes so great that creatures of our cognitive and computational abilities cannot move from models and approximations to the nomological generalizations governing the biological processes. Accordingly, we have recourse to heuristic devices that enable us to do the best we can. If we were much smarter, we would not need these instruments but would be able to employ the very complex general laws that do govern these complex phenomena. If we were very much smarter, biological theory would be very different, but physical and chemical theory would not be.

As noted above, my thesis does not require that we choose between realism and instrumentalism *tout court*. Realists may interpret my thesis as the claim that physics and chemistry should be given a realistic interpretation, while biology is given an instrumental one for reasons acceptable to the realist. After all, no realist has any objection to the use in a realistically interpreted science of heuristic devices, useful fiction, unrealistic models, abstractions, and idealizations. These are necessary tools for advancing our understanding of the way the world is. The realist's distinctive view is that, aside from these artifices, there is a way the world is independent of our theorizing, whether we are clever enough to limn that way or not. When for some reason we turn out not to be clever enough, the realist should have no objection to the resort to instruments useful to us.

Instrumentalists about science in general should treat my argument as a further and separate argument for the instrumentalism of some sciences, one that does not apply to the physical sciences. This further argument may reinforce their much more general arguments from fundamental philosophical premises for all-around instrumentalism. Biology will turn out to be instrumental for more reasons than physical science is.

In principle the dispute between realism and instrumentalism should be neutral on the question of whether nature is simple enough for us to discover regularities about it. If we can discover such regularities, the instrumentalist and the realist will fall to disputing whether such regularities provide knowledge beyond the sequence of sensory data to which we are subjected. My thesis, however, is not neutral on the simplicity of nature; it claims that nature is sufficiently complicated that *we* cannot hope to discover the regularities that operate at the level of the biological. Thus in biology we must content ourselves with heuristic devices, useful instruments.

Having distinguished realism and instrumentalism and having employed this distinction to delineate my thesis about biology, I assume for the purposes of the rest of this work the truth of realism about chemistry and physics. The argument about biology's relatively instrumental character proceeds more simply on the assumption that the physical sciences are not instrumental. Hereafter the argument assumes that findings from physics and chemistry must be interpreted as true, or at least successively approximating to truth.

Now my thesis should be a bit clearer. Biology is an instrumental science relative to a realist's interpretation of physics and chemistry. This thesis should be of interest and significance independent of the broader and more purely philosophical dispute between realists and antirealists (a more fashionable term for instrumentalists these days). In fact, it should shed light on several of the most vexing questions in the philosophy of biology. To begin with, an appreciation of the instrumental character of biological theory will help explain why its relations to physical science are so intricate and why it is so much more autonomous a science in its concepts and theories. The instrumental character of the discipline rationalizes and explains the centrality that models play in the theory even in the absence of well-articulated theory. We will also be able to identify and explain more clearly the limitations on generality and precision in explanation and prediction that biological science faces. And we will be able to retain a clear commitment to the unity of the sciences.

THE UNITY OF SCIENCE

In the period since the lapse of logical empiricism's hegemony in the philosophy of science, the philosophy of biology has flourished. One reason it has is the interest that revolutionary developments in molecular and evolutionary biology have generated among philosophers brought up on a one-sided diet of physical theory. But the philosophy of biology has found a receptive audience because of the apparent differences between the methods of biologists and the characteristics of biological theory and those of physicists and physics. These differences have enabled biologists and philosophers to beat the dead horse of logical empiricism or positivism and in particular to deny the thesis of the unity of science central to logical positivism.

Though most of the rest of the distinctively positivist doctrines have been pretty convincingly disposed of, largely by the positivists and their students themselves,[3] these philosophers never surrendered their commitment to the thesis of the unity of science. The thesis comes in two, perhaps separable, parts: a claim about the methods and epistemology of science, and a substantive or metaphysical claim about the relation of scientific theories to one another.

The methodological doctrine of the unity of science states that all sciences adopt what is broadly the same means of justifying scientific claims, that is, the method of controlled empirical or observational inquiry, with the premium it puts on prediction; the same strategy for explanation, that is, the appeal to regularities or laws with predictive power to account for the occurrence of particular events; the same

3. See, for instance, the papers on the positivist criterion of cognitive significance in Carl Hempel, *Aspects of Scientific Explanation and Other Essays* (New York: Free Press, 1965), and on the analytic/synthetic distinction in empiricist epistemology in Willard V. O. Quine, *From a Logical Point of View* (Cambridge: Harvard University Press, 1953). These works are the locus classicus of positivism's downfall among philosophers, not the writings of Kuhn or Feyerabend, still less their followers in the sociology of science.

theoretical structure in which less fundamental laws are explained by derivation from more fundamental ones. Positivists and their successors spent much effort making these notions more precise and attempting to show how they are reflected in the practice of scientists—mainly physicists and chemists.

The substantive or metaphysical claim of the unity of science is one positivists themselves refused openly to make, but their successors among the scientific realists and materialists were less reticent. The claim is that nature is at bottom both simple and regular, so that we should expect well-confirmed theories in the several sciences to bear an explanatory relation to one another, which reflects this fundamental simplicity and regularity. That is, physical theory deals with the most fundamental and ubiquitous forces and substances, and therefore, since chemical theory describes the behavior of physical substance, its regularities should be derivable from those governing physical forces and substances. Similarly, biological phenomena reflect simply a further level of organization of physical substance, and its theories should be explained by derivation from more fundamental theory in chemistry and physics. And so on for psychology, and ultimately the rest of social science. This doctrine of the substantive unity of science is perhaps better known under the rubric of *reductionism*.

The reductionist's picture, reminiscent of Auguste Comte's nineteenth-century positivism, was not one logical positivists wished to advertise. In part because it made a substantive claim about matters of empirical fact, this doctrine was potentially embarrassing to a philosophical movement that accorded all rights over the establishment of empirical facts to the sciences, not to philosophy. In part the doctrine embarrassed positivists because, until all scientific theories reflect this deductive hierarchy, the only basis on which to endorse it is metaphysical materialism, and positivism was the scourge of metaphysics.

But the empiricist successors of the logical positivists were not as reluctant to make substantive claims about the future structure of scientific theories on the basis of a metaphysical conviction. To begin with, from the 1950s onward, the distinction between the empirical or factual and the nonempirical or logical became too cloudy to serve any longer as a demarcation line beyond which philosophers could not go. Second, the positivists' attempt to devise a litmus test for distinguishing cognitively meaningless metaphysics from theoretical science had been surrendered as a will-o'-the-wisp. Accordingly, the thesis, which positivists and their followers all believed, that societies were nothing but individual human beings, that human beings, like all other living things, were nothing but complex biochemical structures, and that biochemical structures were just physical ones, became increasingly acceptable to endorse publicly.

The metaphysical thesis of scientific unity and the epistemological thesis of empirical unity are clearly well matched; indeed, it might well be said that they are made for each other. Empiricist methods have over the course of several centuries produced a sequence of ever more powerful materialist theories about matter at various levels of aggregation, in which the properties manifested at each level are reductively explained by appeal to more fundamental ones, until we reach the level of the quantum mechanical. And metaphysical materialism has provided the assurance that there are no facts beyond the ken of an empiricist method; there are no things, events, properties, relations, or processes that cannot be discovered and systematized by a scientific method that limits itself to observationally controlled inquiry. Immaterial interventions within the course of nature are just the sorts of things empirical methods should lead us to reject, for, being immaterial, they are not open to detection. Indeed, the more convinced one is of the operation of nonmaterial forces, the less confidence one will repose in empirical methods. Materialism as a metaphysical commitment does not sit well with theories otherwise well confirmed in which nonmaterial forces figure. Thus materialist discomfort with gravitation as a force acting instantaneously through a vacuum motivated the empirical inquiry that resulted in the general theory of relativity and its materialist account of gravitation as the curvature of space.

If nature is not after all regular in the way materialism requires, then there is a range of events and processes that cannot be explained by methods that reflect the unity of science. Indeed, without regularities there can be no certification of knowledge through the confirmation of predictions, for prediction would not be possible. If properties of matter at one level of the aggregation of matter do not consist in properties of matter at lower levels of aggregation, then we cannot expect the sort of derivational relation between derived and fundamental regularities that the unity of science requires theory to consist in. The writ of empiricist methods does not extend beyond the territory of materialism. Without reduction and regularity, empiricist strictures on scientific method cannot commend themselves in the face of competitors. Without regularities, either scientific knowledge will cease or the prediction of empirical observations will no longer be the sine qua non of scientific knowledge. Without reduction of properties and regularities ultimately to those countenanced in physics, there will be explanatory gaps that call for methods to fill them that do not, like empiricism, rely on theoretical explanation by unification.

The gravest threat to the thesis of the unity of science is manifest in the character of biological theory. Along almost every dimension on which we might measure biology's coherence with the content and structure of physical science, it diverges from these disciplines. Biology's divergence from the unity of science is a far more powerful argument against this unity than any merely philosophical asseveration. After all, if a philosophy of science fails to be reflected in the working of real science, so much the worse for the philosophy.

THE DISUNITY OF BIOLOGY

The consequences of methodological and substantive disunity of the sciences are potentially serious, immediately for the philosophy of science and epistemology, and eventually for standards by which we adjudicate scientific theories, explanations, and applications. In a work that came into my hands after much of this book had been written, John Dupré presents a particularly clear version of the argument from biology against the unity of science and an illustration of its potential consequences.

> [Dupré's] book [*The Disorder of Things*] has two interwoven theses. The first concerns science. It is the denial that science constitutes, or ever could constitute, a single, unified project. The second is metaphysical, a thesis about how the world is. This thesis is an assertion of the extreme diversity of the contents of the world. There are countless kinds of things, I maintain, subject each to its own characteristic behavior and interactions. In addition, I propose a relation between these two theses: the second shows the inevitability of the first.[4]

It is worth noting that in effect Dupré endorses the claim that regularity and reduction underlie the prospects for a unified science, though his

---

4. *The Disorder of Things* (Cambridge: Harvard University Press, 1993), p. 1.

interest is in using this relation in a *modus tollens* argument against the unity of science.[5]

Dupré's argument against the unity of science proceeds from science itself—from biology.

> I place myself firmly in the philosophical tradition that sees empirical, often scientific, inquiry as providing the most credible source of knowledge of how things are. In contrast to most related endeavors, however, I shall draw primarily not on physics, but on biology. Biology is surely the science that addresses much of what is of greatest concern to us biological beings, and if it cannot serve as a paradigm for science, then science is a far less interesting undertaking than is generally supposed. (p. 1)

Dupré notes that "the metaphysics of modern science . . . posit[s] a deterministic, fully law-governed, and potentially fully intelligible structure that pervades the material universe." That there are exceptionless general laws, that there is a single order of objective kinds into which the furniture of the universe can correctly be classified and related by general theories, that regularities displayed by kinds at one level of aggregation can be reduced to regularities at more fundamental levels (reductionism)—these are all doctrines Dupré takes to be called into question by the actual character of biological science. "The dream of an ultimate and unified science is a mere pipe dream."

It is remarkable that Dupré explicitly rejects the view I defend here as a reasonable implication of our inability to substantiate the metaphysics of science in modern biology: "my thesis will be that the disunity of science is not merely an unfortunate consequence of our limited computational or other cognitive capacities, but rather reflects accurately the underlying ontological complexity of the world, the disorder of things" (p. 7). Dupré also finds important morals for epistemological diversity to draw from the metaphysical disunity of the sciences.

> [R]ejecting the possibility of any single criterion that constitutes a body of belief as scientific and hence epistemologically acceptable, I also advocate epistemological pluralism. . . . I would certainly reject the dogmatic monotheism of much contemporary

5. The connection is explicitly made in an argument that causal determinism entails reduction: the failure of reductionism in biology "can be seen as preparing the inversion of the reductionist *modus ponens* (causal completeness requires reductionism) into my antireductionist *modus tollens* (the failure of reductionism implies the falsity of causal completeness)" (ibid., pp. 101–102; see pp. 99–102).

philosophy of science; there are surely paths to knowledge very different from those currently sanctioned. . . . We should attempt to develop a catalogue of epistemic virtues. Some of these will flow naturally from the philosophical tradition, consistency, . . . elegance, . . . simplicity. . . . More recent conceptions of science suggest that these must be supplemented with more straightforwardly normative virtues. Investigation of the androcentric and ethnocentric biases of much science suggests, for example, a fundamental desideratum of democratic inclusion and accountability. (p. 13)

If biology is in fact testimony to the disunity of science, as Dupré holds, the ramifications for epistemology are indeed substantial. It is not just that biology, on his view, teaches us the importance of nonsexist, multicultural epistemology, but that it also accords a role to sociopolitical standards as well as epistemic standards in the assessment of good science (p. 13).

I believe that Dupré is right to think that the wider moral of any successful demonstration that biology undercuts the doctrine of unity of science is a substantial change in epistemology in the direction of greater pluralism. Indeed, the consequence seems likely to me to be the most significant single slide down the slippery slope toward the sort of epistemic anarchism characteristic of much of the philosophy of social science.[6]

Epistemological pluralism is the thesis that there are many different ways in which belief can be certified as knowledge, that there are many different kinds of knowledge. Epistemological pluralism raises the question of what gives disciplines with a diversity of epistemologies the right to claim that their epistemologies are all recipes for the same thing, the certification of *knowledge*. If they are so different from one another and share nothing nontrivial, there is no reason to believe they provide noncompeting alternative accounts of the same thing, knowledge. If the term 'knowledge' is merely honorific or meretricious, we might be satisfied with the reply that they need have nothing in common but the label, or that perhaps there is a "family resemblance" among the epistemologies of the varying disciplines claiming to provide knowledge (to invoke a fashionable notion). But this seems to me inevitably to reduce to second-class citizenship those disciplines that do not satisfy the epistemology of the physical sciences. If biology is such a discipline,

6. See, for example, Donald McCloskey, *The Rhetoric of Economics* (Madison: University of Wisconsin Press, 1985).

the ultimate conclusion of an argument from biology for epistemological pluralism is biology's consignment to this second-class citizenship, as a "soft" science.

INSTRUMENTAL DISUNITY

Biology is not a "soft" science, and therefore the unity-of-science thesis and biology need to be reconciled. Failure to reconcile them undermines each. At present it is probably more important to the viability of the unity-of-science doctrine that it be reconciled with the actual character of biology, for the achievements of biology since Darwin have made it a paradigm science, and any philosophy of science at odds with biology must be cast into doubt. On the other hand, it is important for those disciplines that have not yet produced achievements comparable to biology's—the social and behavioral sciences—that the ideal of the unity of science not be cast down. Without it they are condemned to internecine strife between those whose findings and theories satisfy standards drawn from the "hard" sciences and those whose claims cannot meet such standards and who therefore seek reasons to reject the standards.

The claim that biology is a *relatively instrumental* science can provide the required reconciliation and so vindicate a reasonably strong version of the thesis of the unity of science. It will show why much of biology is not smoothly reducible to the physical sciences and, by a natural extension, why psychology is not smoothly reducible to neuroscience, and why behavioral and social sciences cannot be expected to reduce smoothly to psychology. The thesis of relative instrumentalism can explain this irreducibility, even though there are neither differences in the epistemologies of these disciplines, nor differences in their subject matter, nor emergent differences in the levels at which their phenomena are organized.

In chapter 2 I provide distinctively biological reasons for concluding that it is in the nature of biological processes, particularly the mechanisms of natural selection, that biology is an instrumental science. At the level of organization at which natural selection intervenes and begins to channel natural developments, phenomena become so complex that a full account of them passes beyond our computational and cognitive powers. Accordingly, any account of biological processes that does not transcend our powers will be justified not on the adequacy of its descriptions, but on its ability to meet the needs and interests of cognitive agents like us. In effect, this chapter provides a realist explanation for why biology is an instrumental science. In chapter 3 I examine the implications of my biological explanation regarding the complexity of biological processes for a recent powerful account of the nonreductive relations among biological theories.

Chapter 4 extends this approach to our understanding of a key component of evolutionary theory, drift. It argues that the kind of probabilities reflected in this statistical notion must be subjective, or Bayesian, and cannot reflect indeterminacies in the world as do those of quantum mechanics. If drift is an indispensable component of evolutionary biology, then the theory of natural selection is irretrievably subjective. By this I do not mean that it is doubtful or partial or biased. Rather, I mean that it contains an irreducible place for a cognitive agent. It is about flora and fauna, *and* about cognitive agents who theorize about them. In this respect biological theory will perforce differ from physical and chemical theories, which have no such place for the physicist and the chemist.[7]

In chapter 5 I argue that the very levels of organization countenanced by biology reflect our interests. They do not carve nature at the joints but break it up at places that reflect human needs—the need to control our environment in order to secure food, fiber, health, amusement, and so forth. The debate about the levels at which selection operates—the group, the individual, the genotype, the gene—has been among the liveliest among biologists and philosophers of biology. Armed with what the previous chapters reveal about biological theory and its relation to the rest of natural science, we can put in perspective the dispute about whether the units convenient for biological theorizing are also those to which we should be ontologically committed. If I am correct about the nature of biological theory, then even on the standards established by opponents of the thesis that the gene is the only unit of selection, they must themselves eventually accept this conclusion.

Chapter 6 examines whether the claim that the rest of biological theorizing is instrumental also applies to the theory of natural selection. This question is crucial for my strategy, for the argument that biology is an instrumental science rests mainly on the fact that biological processes are the result of the operation of random variation and natural

7. Except perhaps in the Copenhagen interpretation of quantum mechanics. But we are assuming that instrumentalism in physics is false. Even if physics were instrumental, as the Copenhagen interpretation might suggest, by itself this will not allow for subjectivity at the level of biological organization. See chapter 3.

selection. If our commitment to the theory of natural selection is itself in part heuristic, if the theory of evolution is itself a useful fiction for creatures no smarter than we are, then the linchpin of my argument is pulled away. After all, my strategy is to provide realist arguments for biological instrumentalism. I cannot afford to include as a premise a theory that itself must be interpreted instrumentally, lest the argument beg the question of why biology is an instrumental science. Accordingly, I try to show that, unlike the rest of biology, the theory of natural selection does not reflect a complex combination of facts about the world and about us, but is unambiguously about a process in which we play no role.

The chapter goes on to explain why, nevertheless, philosophers of biology have been so strongly attracted to a view of the theory that makes it a body of models instead of a set of laws operating everywhere and always quite independent of us. Philosophers have embraced this view because they mistake the pure theory of natural selection for the varied contexts in our world to which we apply it. Since, applied to our world, the theory is combined with components whose role is purely heuristic, the theory is itself wrongly so viewed.

In chapters 7 and 8 I extend the strategy of the analysis of biology to psychological theorizing and to the rest of the behavioral and social sciences. If biology is an instrumental science, then so is every science about biological creatures like us. Chapter 7 examines attempts to understand and to vindicate popular forms of psychological theory by drawing parallels between them and biological theory. I argue that the parallels can at most show that psychological theory is, like biological theory, a body of heuristic devices and not an autonomous discipline uncovering its own distinctive laws and theories. This finding would not be so threatening to popular views about the character of psychological theory, were it not for the fact that it undermines these theories' claims to exclusivity in the attempt to account for human behavior. If some particular psychological theory on the one hand is not particularly predictively powerful and on the other hand can be justified only by its heuristic role, then we have every incentive to search for a more useful instrument. This, I hold, is the status of intentional psychology, the theory that attributes to human beings and other animals psychological states with semantic content.

In chapter 8 I summarize the implications of my argument for biology and show that conclusions about the character of psychological theory must also temper our expectations in behavioral science.

Whatever Happened to Reductionism, and Why?

UNTIL THE DISCOVERIES of Watson and Crick, biological reductionism was mere philosophy. From the philosophical reception of their discoveries in the early sixties until the age of genetic engineering, reductionism was gospel. Now, like 'positivism' it is a term of derision and abuse for a doctrine widely repudiated in the philosophy of biology.[1]

Complete reduction is what happens when the generalizations of one theory are shown to be explainable by the generalizations of a more fundamental one, without leaving out anything crucial captured by the narrower theory. For about fifteen years after Watson and Crick were awarded the Nobel prize (1962), the prospects seemed bright that in biology we could find what all recognized to obtain in the physical sciences: relatively "smooth" reductions of less fundamental theories to more fundamental ones, both across the history of science and down the hierarchy of completed scientific achievements. The long history in physics and chemistry reveals increasingly close approximations to complete reduction of less basic theories to more basic ones, of less general theories to more general ones. This history reflected the concrete vindication of the positivist's doctrine of the unity of science. It is also the best argument for the existence of a small number of fundamental physical kinds, and laws of working that govern their behavior, underlying and fixing the apparent buzzing, blooming confusion of nature as we observe it. The conviction that so many diverse physical and chemical processes can be systematized and explained by showing

---

1. Sometimes the term 'reductionism' is employed as a term of abuse for theories that claim to reduce the number of causal factors otherwise supposed to be relevant for explaining some process. Thus some sociobiologists are sometimes accused of reductionism because they apparently deny the causal relevance of environmental factors, along with inherited ones, in the explanation of behavior. This is not the sense of reduction or reductionism at issue.

that the appearance of ceaseless change is underlaid by continuity, conservation, and changelessness rests on nothing more or less than this history of the last four hundred years of science.

Of course, everyone recognized that the simple account of reduction, in terms of derivability of laws and connectability of terms, was too coarse. Nevertheless, something about the relations between Kepler's laws and Galileo's, on the one hand, and Newton's, on the other, substantiated the reductionist's picture. The more complex cases of theoretical subsumption that followed down to quantum electromagnetics and the general theory of relativity seem to lend equal support to the doctrine. At least since the scientific revolution of the seventeenth century, the history of scientific progress and cumulation is in large measure a history of the reduction of the general to the more general.

Some recalcitrant problems needed to be dealt with in order to substantiate fully this picture of the simplicity of nature. For example, there were problems about connecting terms with incompatible "meanings" like Einsteinian 'mass' and Newtonian 'mass', which the simple reductionistic picture unintelligibly equated. The former is a relational property whose quantity varies with reference frames, while the latter is a monadic property intrinsic to the object and independent of any other state of affairs. How could the two terms be equated, as seems to be required for the deductive derivation of Newton's laws from the special theory of relativity and the assumption that the speed of light is infinite? No immediately acceptable conclusion was forthcoming. But these were presumably problems in the philosophy of language, problems about meaning and reference, to which the philosophy of science could assume a solution that would preserve the reductionist picture.

Then, as Schaffner showed, reduced theories need to be "corrected."[2] And there were problems associated with what counts as a correction of the reduced theory, as opposed to a new theory altogether. But these seemed to be problems about systematizing intuitions that were pretty clear and strong enough to overrule the problems.

Reduction certainly seemed to be the rule and not the exception in the history of science. By and large the philosophy of biology was slow to adopt the radical claims of Kuhn and Feyerabend about the incommensurability of theories and the irreducibility even of Kepler's

2. See Kenneth Schaffner, "Approaches to Reduction," *Philosophy of Science* 34 (1967): 137–147.

and Galileo's discoveries to Newton's laws, let alone the irreducibility of Newtonian mechanics to the general theory of relativity. In the philosophy of biology the decision about reducibility was made on the merits of the case in biology and not as another inference from events in physics and its philosophy. In fact, as Ruse confidently assured the readers of *The Philosophy of Biology* in 1973, the evidence substantiating a smooth reduction of Mendelian genetics to molecular genetics was coming in rapidly. Reducing Mendelian genetics to molecular genetics was widely thought tantamount to reducing theories about living things to theories about inanimate phenomena. Once accomplished in genetics, the cat was out of the bag, the in-principle possibility of a fully physical account of all biological process was as good as proved. Reduction, it turned out, had only awaited biology's getting its house in good enough order to frame precise theories so we could see exactly how they were to be absorbed by chemistry and physics. Not only was the unity of science vindicated at the critical juncture between physical science and life science, but the metaphysical thesis of materialism was given a powerful scientific vindication. Here at last was factual proof, not merely philosophical argument, with which to confront the dualist and the antimaterialist in metaphysics. There is a strong intimation of these philosophical by-products of the discoveries of the early sixties in *Chance and Necessity* by Jacques Monod, who received a Nobel prize for his work in molecular biology of the gene.

THE PROBLEM OF THE MANY AND THE MANY

In the midst of all these expressions of confidence in the movement of biological science toward close articulation with physical science, David Hull began the process of unravelling the scientific and philosophical consensus. In the early seventies he showed that there is another problem, characteristic of the relations between molecular biology and Mendelian genetics, that effectively blocked any actual reduction.[3] Echoing Plato's problem of "the one and the many," let us call it the problem of "the many and the many." Plato's problem was explaining how different things—say, all red things—could share the same property— redness of color. His solution was the existence of a single form of redness in which the many individual red things could "participate."

3. *The Philosophy of Biological Science* (Englewood Cliffs, NJ: Prentice Hall, 1974).

Hull's problem is interestingly similar, though his solution is nothing like Plato's!

The problem of the many and the many is simple enough to state. Here is Hull's own exposition.

> Even if all gross phenotypic traits are translated into molecularly characterized traits, the relation between Mendelian and molecular-characterized predicate terms expresses prohibitively complex, many-many relations. Phenomena characterized by a single Mendelian predicate term can be produced by several different types of molecular mechanisms. Hence, any possible reduction will be complex. Conversely, the same type of molecular mechanism can produce phenomena that must be characterized by different Mendelian predicate terms. Hence reduction is impossible. (*Philosophy of Biological Science*, p. 39)

If anything, the explosion of discoveries in molecular biology since 1974 has shown that matters are even more complicated than Hull supposed. But his discernment of the problem was prescient indeed. For any Mendelian predicate, there is a disjunction of actual alternative molecular mechanisms, and any molecular predicate will be implicated in several different Mendelian processes.[4]

The consequences of this many-many relation between molecular and Mendelian predicates, and the properties they denote, are obvious: if there are nomological generalizations at the level of Mendelian processes—say, the "laws" of segregation and independent assortment, or population genetics—like the Hardy-Weinberg law, these will be *autonomous* from any generalizations at the molecular level. They will not be derivable from any number of molecular general laws, for there is no way to link Mendelian principles to molecular generalizations. The reason is that we cannot frame bridge principles of the sort that

---

4. The many-many problem obtains even if we embrace Waters' important insight that "the operative" genotype/phenotype relation in classical genetics is that of gene difference/phenotypic difference, and not gene/phenotype ("Genes Made Molecular," *Philosophy of Science*, forthcoming). To begin with, the many-many relation obtains between molecular differences and classical gene differences, as well as between molecular phenotype differences and classical ones. Second, and perhaps more important, gene/character differences were and are, as Waters puts it, "operative," because phenotypic data, molecular and nonmolecular, reveals differences; the entities that bear these differences are always inferred to explain how the gene differences make for phenotypic ones. So in the end it is the entities—the classical and the molecular genes—that are the locus of differences and that must be related. See chapter 3, note 2.

reduction requires to connect the predicates of Mendelian genetics to those of molecular genetics.

We may put the matter starkly by using a little of the notation of predicate logic and seriously oversimplifying both Mendelian processes and molecular ones. Suppose first that there is a Mendelian law of the form $(x)(Dx \to Wx)$ for "All things with a dominant gene for whiteness, D, are white, W," and second that there is a molecular generalization of the form $(x)(Nx \to Px)$ for "All things with nucleic acid sequence N express protein P," where the Mendelian generalization is to be derived from the molecular gene. Again, there are no such simple generalizations of either theory, but for the sake of illustrating Hull's point, let's see what is required to derive one from the other. Doing this requires bridge principles of the form $(x)(Nx \leftrightarrow Dx)$ and (x) $(Px \leftrightarrow Wx)$. The problem is that there are no such bridge principles. At most we will find biconditionals that are "disjunctive" and incomplete: expressions like

$$(x)(Nx \leftrightarrow Dx \vee Fx \vee Gx \vee \ldots),$$

where properties F, G, and so forth, are other Mendelian properties, and the ellipses indicate that there are other unknown Mendelian properties in the appearance of which the molecular property N is implicated. Similarly, the biconditional connecting the protein P and the Mendelian property of whiteness W will have the same disjunctive structure:

$$(x)(Px \leftrightarrow Wx \vee Sx \vee Tx \vee \ldots).$$

It is already clear that, if the bridge principles take this form, no derivation is possible. The most we can deduce from our molecular generalization and these bridge principles is a disjunctive generalization of the form

$$(x)(Dx \vee Fx \vee Gx \vee \ldots \to Wx \vee Tx \vee Sx \vee \ldots).$$

And a generalization with a vast disjunctive antecedent, an equally vast disjunctive consequent, and ellipses will not pass muster as an interesting generalization or even the first approximation to one.

Matters are in fact a good deal worse, since even these bridge principles seriously falsify the relation between molecular properties and Mendelian ones. Any single Mendelian property is the result of the occurrence of a hideously complex conjunction and disjunction of molecular properties. In fact the bridge principles, if we bothered constructing them, would be roughly of the form of an indefinitely long biconditional, each side of which is a disjunction over a vast number of further disjunctions. It will look like this:

$$(x)\,(Dx \lor Fx \lor Gx \lor \ldots \leftrightarrow Wx \lor Sx \lor Tx \lor \ldots).$$

But any two arbitrarily chosen predicates are probably components in a true statement of this form. It would not take much ingenuity to construct a biconditional of this form for the predicates "x talks like Daffy Duck" and "x has quantized angular momentum." The predicates of Mendelian and molecular theory do not line up in simple enough relations to make any kind of smooth reduction possible. So any strict deductive derivation, from molecular biology, of a Mendelian phenomenon in its generality may implicate much or all of what theory there is about all the macromolecules known to molecular biology. To use our disjunctive bridge principle to go from molecular laws to Mendelian laws requires that we exclude a multitude of inoperative disjuncts.

This sort of reduction is not only methodologically useless, it is probably unattainable by agents of our cognitive and computational powers. Reductionism thus seems fated to cast little light on intertheoretical relations in biology.

These are the facts of life to which materialists and other erstwhile reductionists must reconcile themselves. One way to do so is to expand or change the notion of reduction in order to accommodate the impossibility of deriving generalizations from generalizations alone. Given the roles in science of mathematical deduction and explanatory unification, this is a drastic change. A more reasonable response is to accept that the systematic deduction of theory that characterizes physics and chemistry is not to be had in biology, to explain why not, and to show that the absence of reduction is no threat either to materialism or to the unity of science.

This is where the philosophical concept of supervenience enters. Erstwhile reductionists agree that a given polynucleotide sequence may result in two or more effects recognized in Mendelian genetics, or vice versa, a given Mendelian effect may be the result of two or more different biosynthetic pathways from two or more different polynucleotide sequences. When this occurs, it is because of variations in other causally relevant factors, especially in the molecular milieu of the genetic material in the nucleus and the cell in which gene products are synthesized. Adding these factors to a biochemical-Mendelian biconditional would provide the bridge laws without the ellipses required to effect the reduction; it would convert the many-many relation to a one-one relation of the sort reduction requires. Admittedly, the result would be a biconditional of great length and complexity, and neither the generalizations that such biconditionals could be combined with nor the generalizations they would jointly imply are recognizable components of any biological theory, molecular or not.

But adding these conditions does reflect the assumption of mereological determinism of Mendelian phenomena by molecular phenomena that the unity of science requires. Mendelian properties are *supervenient* on molecular ones: given any two biological systems that are identical in all their molecular properties, they will *have to* be identical in all their Mendelian properties. Two biological systems with different Mendelian properties will have to differ in some molecular property or other, although two biological systems may be identical in Mendelian properties while differing in molecular ones. This latter possibility merely reflects the fact that the one molecular property can be combined with different packages of other molecular properties to realize different Mendelian properties, while the same Mendelian properties may be produced by several different molecular packages. The only alternative to this sort of mereological determinism is to embrace a thesis that the whole is ontologically more than or different from the sum of its parts. This is not a doctrine any exponent of the unity of science can endorse, nor would opponents of the unity of science want to be saddled with it. The history of science and any reasonable epistemology places the burden of proof on such holists to show how the emergent properties of wholes are possible.

In general, if As are supervenient on Bs, then As "are nothing but" Bs, even though two As may differ in the Bs they are made of, so long as two identical Bs are both As. The relation of supervenience can obtain even though *we* can't explain how it is in general that all As are Bs. How is supervenience possible? Suppose we distinguish between two sorts of descriptions: structural and causal description. Roughly, the structural description of a particular item identifies the material it is composed of and its spatially distinct and separate parts. Thus an ax is structurally described as composed of a wooden shaft of oblong cross section between two and three feet in length to which is attached a steel rectangular prism with a sharp edge. A causal description of an item identifies it by citing its causal role, the characteristic causes and effects of its presence in its typical context of appearance. Causal descriptions are often called 'functional' role descriptions in philosophy of science, and I will use this terminology hereafter, understanding 'functional' to mean simply "role in a network of causes and effects." Thus the functional description of an ax identifies it by its functional role, the cases and effects into which it typically enters. An ax is a tool (that is, its existence is brought about as a result of certain human intentions) for cutting wood (that is, its effect when applied with sufficient momentum will be to sever wood into smaller pieces). Of course, most descriptions of objects combine structural and functional description: if I describe an ax as having a wooden handle and a steel head, then by employing the terms 'head' and 'handle,' I advert to its function along with a description of its composition.

Now consider the class of objects that meet the purely functional characterization of ax. They all have to be material objects and have some structural composition or other. But nothing in the functional definition requires that their handles be made of wood or that their heads be made of steel. For many purposes metal or graphite or plastic handles will do as well. Similarly, the head need not be made of any one kind of steel; it may be of carbide steel, stainless steel, galvanized steel, or iron or of aluminum or gold, or it may be made of stone, for that matter. Of course, axes of some material compositions may not be as good in discharging the function of axes as those of other compositions, but they are all recognizably axes.

An ax is thus "nothing but" the material out of which it is composed, even though two different axes may share no fact of structural composition in common (above the atomic level); so long as any two objects are structurally identical and one can function as an ax, then the other one must be able to function as an ax too. Being an ax is thus supervenient on having some disjunction of packages of physical composition, each of which has some minimum set of capacities with respect to our interests in cutting wood.

These facts are common to all classes of functionally defined or described objects. Functional kinds supervene on structural kinds. The relevance to the materialist and erstwhile reductionist's claim is obvious. The terms of Mendelian genetics—gene, dominant, recessive, epistatic, mutation, crossover, linkage, heterozygote, and so forth—are all functional terms, and the natural kinds of Mendelian genetics are all functional kinds. The kinds of molecular biology are structural. Or to be more precise, they are structural *relative* to Mendelian kinds. Relative to the kinds of organic chemistry, they may turn out to be functional descriptions. For example, to call something an enzyme is to identify its function, catalysis, and its structure, a protein. But as molecular biologists well know, being an enzyme is supervenient on a disjunction of different chemical active sites, and being a protein is supervenient on a large number of different primary sequences of amino acids.

So erstwhile reductionists shift from their claim that Mendelian genetics is reducible to molecular genetics to the thesis that the former is supervenient on the latter. Thus, instead of expecting the deductive derivation of general laws of Mendelian biology from more general laws of molecular biology, we should expect at most the identification of the molecular mechanisms that subserve relatively small classes of Mendelian phenomena on a case-by-case basis. We should expect the identification of interesting commonalities and differences among the disjunction of molecular mechanisms that underlie a given Mendelian phenomena. As for the laws of these two theories, well, if there are laws in both of these domains, they will not be related to one another in any simple way.[5]

Two questions emerge from this concession. First is the substantive biological question, why is reduction of the sort envisioned by logical empiricism impossible in biology? Why should the relatively smooth reduction in chemistry and physics break down in biology? This question is a request for a causal explanation, one that identifies the biological facts that result in the many-many relation. It must be sharply distinguished from the methodological question of the same verbal form (Why is reduction impossible in biology?) whose answer is the problem of the many and the many. The question that needs to be addressed is what *causal* facts about biological systems result in the theoretical problem of the many and the many, when such difficulties do not seem to obtain in physical science. Second is the question, what view of the nature of biological science emerges from reflection on this biological fact and from the denial that reduction obtains at all between these two theories?

SELECTION FOR FUNCTION IS BLIND TO STRUCTURE

What facts about biological systems make "layer-cake" reductionism, as it has sometimes been called, impossible in the life sciences? Biologi-

5. This view of the erstwhile reductionist is exemplified in Alexander Rosenberg, *The Structure of Biological Science* (Cambridge: Cambridge University Press, 1985), chapter 4.

cal systems are the result of natural selection over blind variation. The fact that they are the result of adaptational evolution is the causal fact that explains the impossibility of reduction. In retrospect the explanation should not be surprising, since evolutionary adaptation is what is distinctive about these systems, by contrast to purely physical ones. So it seems the obvious first place to look for a causal explanation of the differences between theories about biological systems and theories about purely physical systems. In particular, such an explanation seems vastly to be preferred to one in terms of "paradigm shift" or Feyerabendian incommensurability, or differing social constructions of science. A biological explanation for the impossibility of reduction is simpler, is based on far more secure premises, and is far more precise in its explanatory focus, than any of these nonbiological explanations could be.

What is it about natural selection over variation that precludes reduction by derivation between biological theories? Natural selection "chooses" variants by some of their effects, those we identify as their functions. Processes that are "random" with respect to adaptation result in combinations of many sorts: thus quantum electrodynamic processes result in nuclear and atomic phenomena, chemical bonding processes result in molecules, thermodynamic processes acting together with these other processes result in larger arrays, and so forth. These combinations of matter have various properties, some of which enhance the physical stability and persistence of the combinations, others of which render the combinations unstable. Stability and persistence among these physical combinations are obviously "selected for" by thermodynamics. Stable combinations do not require continuing energy inputs and resist entropy.

Some of those stable combinations replicate themselves. These are molecules of certain configurations that, for example, can catalyze the processes that lead to the creation of more molecules of the same configuration, or foster the thermodynamic circumstances conducive to their appearance, or serve as templates on which combinations of the same kind will come together as a result of random drift of molecules. However they replicate, these configurations are even more strongly selected for than those that are merely stable. They are reproduced in larger and larger numbers. The same goes for larger arrays of molecules, all the way up to macroscopic configurations, of the sort detectable by naked eye. This much verges on tautology.

What is not tautology is that among such combinations, at apparently every level above the polynucleotide, *physically distinct* structures are frequently found with some *identical* or nearly identical functional properties, different combinations of different types of atoms and molecules, that are close enough to being equally stable and equally likely for purely physical causes, to foster the appearance of more instances of the kind they instantiate. So far as adaptation is concerned, there are frequently *ties* for first place in the race to be selected. As with many contests, in case of ties, duplicate prizes are awarded. The prizes are increased representation of the selected types in the next "reproductive generation." Thus, if urea and methane are equally stable, and if the presence of a urea or a methane molecule makes for equal increases in the probability that more urea and methane molecules will be produced in their vicinity, both are equally well adapted. And so on up the chain of chemical being.

It is the nature of any mechanism that selects for effects, that *it cannot discriminate between differing structures with identical effects.* Functional equivalence combined with structural difference must in the nature of the case increase as physical combinations become larger and are more physically differentiated from one another. This may all be obvious at the level of the observable phenotype; as Hull points out in *The Philosophy of Biological Science* (40–41), the steel-gray color of a heterozygous Andalusian fowl could result from a variety of molecular situations: "Perhaps a blue-gray pigment is being produced. Perhaps only a black pigment is being produced, but in reduced concentrations. Perhaps both black and white pigments are being produced in equal amounts. These pigments could be distributed evenly throughout the feathers, or gathered together in small patches. In the latter case, the gray appearance would be a function of the visual acuity of the observer." But the blindness of selection to structure is evident at the most elemental level of molecular biology as well.

Below the level of the nucleotide and the polypeptide, intermolecular combinations do not seem to provide two or more different structures with sufficiently similar effects to both be selected by a mechanism blind with respect to structure. The information that controls development and heredity for all organisms is carried in nucleic acid and only nucleic acid. No other molecule serves as the carrier of hereditary information. That this crucial biological function should be subserved by only *one kind of physical structure* has been the object of a fair amount of speculation among biologists. Almost every other biologically interesting function is subserved by two or more—usually many more—

physically distinct structures and mechanisms. Why should this particularly complicated function be subserved by only one physical structure? Only two answers to this question seem plausible. Nucleic acid is so much better at information storage than any other molecular configuration that in the long period of evolution it beat out all the competition; in the end no duplicate prizes were awarded, and so only organisms bearing DNA or RNA as the genetic material survived. Or perhaps at the level of molecular information carrying, there was only one way nature could skin the cat: given purely physical constraints, the only configuration that was ever capable of bearing information with the fidelity required for transmission and regulation is nucleic acid. Thus on this implausible scenario it was selected for in the degenerate case: there were no competitors to beat out.

Whether or not nucleic acid was selected for as the sole winner in an evolutionary competition, at almost every higher level of configuration that includes nucleic acid as a structural component in the race for selection, there have been ties at the (always local) finish line, and duplicate prizes have been awarded. We may illustrate how the diversity of structures underlies a given function at the very next level of structural organization above the single DNA nucleotide—the triple of nucleic acids that serves as the minimum unit of hereditary information.

The DNA carries information about the order of amino acids which compose proteins. The information is conveyed to the ribosomes by messenger RNAs, which copy the sequence of the DNA. At the ribosome, transfer RNA attaches amino acids to each other in the order determined by the messenger RNA to produce the required protein. But there is "slack" in the genetic code by which DNA carries the original directions for the composition of messenger RNA and eventually proteins in the ribosomes. The polynucleotide chains of DNA need to code for proteins constructed out of sequences of twenty different types of amino acid. The four nucleic acid bases from which the DNA chains are composed, thymine (T), cytosine (C), guanine (G), and adenine (A), can be combined in $4 \times 4 \times 4$ or sixty-four different sequences of three bases each. These triples of nucleic acids are called codons. Since there are only twenty amino acids, only twenty out of the sixty-four codon triples are required (plus sequences that signal termination of amino acid chains which compose proteins). Thus there are enough codon combinations of three bases each to allow for redundancy in the genetic code. Every distinct amino acid besides methionine and tryptophan is coded for by at least two different codon sequences, some by as many as four. Thus valine is coded for by codons with the following sequences: GUU, GUC, GUA, GUG. Any one of these triples signals valine in the DNA's instructions about which amino acid to add next in a protein under construction at the ribosome. Coding for valine is the effect of a disjunction of four alternative sequences of nucleic acids.

One obvious explanation for this redundancy of the genetic code or, as biologists put it, its degeneracy is the absence of a selective difference, either for themselves or for organisms containing them, among these four sequences. If they all code equally well for valine and coding is their *only* functional role, then selection will be blind to their structural differences. But it has perhaps not escaped your notice that all four of these nucleotide sequences—GUU, GUC, GUA, GUG—share the same initial two nucleotides, and differ only in the third position. This in fact is true for all of the types of codons synonymous for a given amino acid.

Now, in molecular biology such structural regularities are always an invitation to function finding. This is what distinguishes molecular biology from organic chemistry. The fact that redundancy appears only at the third position in a codon has led biologists to consider whether the persistence of synonymous codons has an adaptive explanation after all. Indeed, at least one such an adaptive explanation has been offered. If it is right, the redundancy of the DNA code is not the direct result of a tie among DNA triples' structure in a selective race to best convey information, in which duplicate prizes have been awarded. Rather, the redundancy of the genetic code is the result of blindness of selection for function to structural differences among RNAs that operate in the ribosomes several steps removed from the DNA. The redundancy of the DNA code is the indirect result of this blindness of selection at a higher level of aggregation of matter.

Transfer RNAs are the vehicles that bring amino acids to the ribosome, where the messenger RNA, transcribed from the DNA, directs the order in which the amino acids are bonded together into a protein. Each transfer amino acid bears an "anticodon" triple that matches up with the amino acid it bears and bonds to the codon triple transcribed from the nuclear DNA (by the messenger RNA). But when the third nucleotide of the transfer RNA anticodon is uracil (U), the molecular bond linking the transfer RNA to the messenger RNA will be equally strong between the transfer RNA's uracil (U) and the messenger RNA's guanine (G) or adenine (A). Thus, a messenger codon of GUG or GUA will effect the same amino acid bonding by a transfer RNA's CAU anticodon. Whence the redundancy of the genetic code.

Apparently selection has permitted several different molecular sequences to serve as the transfer RNA for a given amino acid. This is what ultimately results in the "degeneracy" of the DNA code. Diversity of structures among transfer RNAs with the same function in protein synthesis permits messenger RNAs to perform the same function while diverging in structure, which in turn allows the DNA the same latitude. Thus selection for function at the level of the transfer RNA transitively allows for diversity of structure at the DNA level.

Why is the liberty taken? Why does DNA structure vary when nothing is forcing it to do so? Perhaps the answer is random variation. When selective constraints are broad enough, structural differences will arise for purely physical causes and are labeled "random drift" in biology. Or perhaps there is an adaptational explanation for why the DNA takes advantage of the opportunity to be redundant: for example, a redundant code is less susceptible to point mutations.

Thus we end up explaining why selection *seems* blind to a structural difference between DNA sequences by showing that it is not in fact blind to them directly, but blind to a structural difference between transfer RNAs. It is this diversity of RNAs with one function that permits the existence of a structural diversity of DNAs with another function.

Sometimes, however, even above the level of the polynucleotide, nature really seems to select for single structures to fulfill a given function. Histones are proteins around which DNA is wound. There are several different histones. Over the last 80 million years, or at least since the lineages of rodents and humans diverged, the gene for histone 3 has apparently experienced no (nonsynonymous) substitutions in its primary sequence, and the amino acid sequence for histone 3 proteins has remained the same among all mammalian species that evolved from the common ancestor or rodents and humans.

The cause of this singular uniqueness of structure across so vast a range of organisms is relatively obvious. Since the amino acids in the histone 3 protein interact directly with DNA bases to support their functions, any substitution in the sequence of the histone 3 protein is likely to impede DNA function. In particular, histone 3 must retain a compact secondary structure (shape) and high alkalinity to interact with the acidic DNA double helix wound around it and the other histones. Mutatis mutandis, for the gene that codes for the histone 3 protein, or at least for the part of the DNA that determines the protein's primary structure.[6] Its structure is narrowly constrained by selection for the function of the protein it produces. In this case we can fill in the bioconditional: a protein is histone 3 if and only if it has one of the following primary structures . . . . (where the disjuncts differ only in synonymous nucleotide substitutions).

These examples present, at a very "low level" of organization, a phenomenon widely cited by opponents of reductionism. That is, the direction of mereological causation seems the reverse of what reductionism requires: transfer RNA structure and function determines DNA structure; histone function fixes DNA sequence. Instead of the DNA structure's determining the character of gene *products,* the molecular structure of the genetic material is explained by these products' organization and function. Nature selects for function in the light of environmental constraints. These constraints include the molecular environment of the genetic material in the nucleus, and the cellular environment, including the organelles operating within the cell. So at least sometimes a given *single* molecular structure may be selected for, because it is optimal in the light of a diversity of physically different systems with the same function. But it is more likely that the physical diversity of these "higher-level" systems will call forth not one but a *diversity* of physically different molecular structures with the same function.

Furthermore, if selection can operate in ways that discriminate between very slight differences in functional efficiency among molecular configurations, then structural diversity among systems with some of the same functions will be encouraged. A given physical system may have indefinitely many effects on its environment. Only a subset of these actually are functions, that is, properties selected for by the systems' environment. And among these functions only a smaller subset will be identified as such by scientific inquiry. Moreover, a physical system may functionally subserve several different "needs" of the large system in which it operates. Thus, for example, thymine is present in DNA to code for amino acids like valine but also to ensure maximum fidelity, since it is less prone to point mutation than the uracil that replaces it,

6. For a discussion see Wen-hsiung Li and Dan Graur, *Fundamentals of Molecular Evolution* (Sunderland, MA: Sinauer, 1991), pp. 67–77.

at lower energetic cost, in RNA. All the diverse functions a physical system has must be reconciled, to the extent it is successfully selected for.

Several structures can fulfill the same function. Moreover, different physical structures can fulfill a bundle of functions with equal "overall" adaptive efficiency. And they may do so by fulfilling different component functions within a bundle differently. Thus some parts of the genetic material may be better at fidelity and worse at energetic efficiency than others, yet, given trade-offs, both sorts may be equal in "overall" adaptation. These differences in individual functions must be traced back to differences in structure, differences that cancel out so far as overall fitness is concerned.

So not only does nature select diverse physical systems because of their identity of function, but the diversity of these selected systems itself also acts as a selective force producing further diversity of structure with identity of function. And of course, it isn't *identity* of function that's required. *Similarity* will do. Indeed, it may be the case that no two physically diverse structures have exactly the same set of functions, for their diversity entails that they have diverse effects. Even if some or all of these diverse effects are inconsequential for selection in any given environment, there is always some possible environment and some length of time in which the smallest structural difference or its immediate effect can bear a selective advantage or disadvantage.

In fact, it is not necessary to come in first at any given local selective finish line in order to receive the prize—the opportunity to replicate further. Because a sufficient number of nonselective forces impinge on configurations and affect their prospects of reproduction from occasion to occasion, just coming a close second or third to being most well adapted is often enough to ensure persistence and replication. This, after all, is what the phenomenon of evolutionary drift is all about. Sometimes, as between two or more variants, the most well adapted is eliminated through purely nonselective accidents. Sometimes the environment does not remain constant for long enough to ensure that the optimally adapted structure becomes fixed or the less optimal structures are completely extinguished. This gives configurations in second place another chance. And if the chances of a potential first-place configuration—once eliminated by nonselective causes, recurring later, and joining the race again—are low enough, these second-place configurations will never have to face competition from it. Or again, when the previously most well adapted configuration recurs, the environment may have changed, or the previously second-place configurations' successors may have been improved by further selection.

Natural selection thus makes functional equivalence cum structural diversity the rule and not the exception. By contrast, nonevolutionary processes (mechanical, thermodynamic, electromagnetic, chemical) make structural difference with equivalent effects the exception, if they permit it at all. About the only example of structural diversity with equivalence of effects in the physical sciences that has been offered in the literature is the various processes on which changes in temperature supervene: change in kinetic energy, change in electrical resistance, change in pressure (holding volume constant), change in length of a material, and so forth. Interestingly, each of these relations has been exploited to serve a human purpose—to design thermometers, instruments that measure temperature change. By contrast to the biological case, however, the total number of these structurally different, functionally equivalent processes is small. More important, every such thermometer shares a single common underlying mechanism at an only slightly lower level of structural organization—the quantum mechanical.

This difference from biology, resulting from the operation of selection for effects, explains why reduction goes smoothly in the physical sciences, and apparently not at all in biology and in every other discipline in which phenomena are explained by their effects.

FROM METAPHYSICS TO METHODOLOGY

Reductionism is in some large measure the reflection of methodological and metaphysical commitments to the uniformity and simplicity of the fundamental constituents out of which the rest of nature is composed. Its best and perhaps only really convincing evidence is to be found in the persistent predictive improvement of physical theory, as its degree of unification under a small number of uniform relations among simple components increases. Reductionism is the attempt to make precise the notion that, under the buzzing, blooming confusion of nature, there is this small number of mechanisms or processes, and a small number of different types of things, which can systematize and explain the world. This notion of simplicity is the cornerstone of reductionism. If the world's apparent complexity cannot ultimately be explained by the simplicity of its base after all, then reductionism's metaphysical rationale and methodological justification is undermined.

What seems to be the moral of natural selection for scientific method and metaphysics is that, above the level of the molecule, nature isn't simple anymore, and it isn't simple because of the blindness to structural differences of selection for functions. Thus, above the level where ties result from nature's selection for function, regularities, such as there are, cannot be explained by a small number of laws about a limited number of physically distinct systems, no matter whether explanation is a matter of logical derivation or some quite different relation.

What is more, the regularities that selection for function generates will be far more restricted, exception-ridden, and complex than we have come to expect in other disciplines. Consider the class of objects selected because they fulfill some adaptive function, Fx. We seek some interesting generalization about items in this class, perhaps something of the form $(x)(Fx \rightarrow Gx)$. Thus we seek another predicate, Gx, true of all items in the extension of Fx. We can seek no structural property of members of the extension of Fx, for the class of Fs is physically heterogeneous, since they have all been selected by effects. Of course, we may find some structural feature shared by many or most of the members of F. But if we do, it will be a relatively uninteresting one, or if interesting, then the class of objects with property F is probably very narrow.

Consider first an example of the former case, the search for some property contingently true of all members of a functionally defined class, say, the class of all aquatic animals. It's easy to identify a property they share in addition to being aquatic animals: they all share the property of experiencing gravitational attraction. But this property is not biologically interesting. It's true, but no law, that all aquatic animals experience gravitational attraction. Clearly, what we want is not just a property necessary for being aquatic but one distinctive of it. This is what we cannot provide. There are thus no interesting generalizations about all aquatic animals. Since functional classes are produced by selection and selection does not discriminate between diverse structures that are equally advantageous with respect to the trait selected for, functional classes are ipso facto heterogeneous, and no further interesting generalizations are true of all their members. Whence it follows that biological laws will be few and far between, for all biological kinds above the level of the macromolecule will be functional.

This brings us to the second example. For a relatively restricted functional property, like the molecular ones, we can produce a disjunctive but relatively manageable class of alternative physical structures about which interesting generalizations are possible. Consider the predicate 'codes for valine'. It names a class composed of instances of three types of DNA sequences, and underwrites the generalization that all DNA sequences that code for valine contain guanine and uracil. This generalization, though exceptionless and precise, is about the best we can do by way of a biological law.

If we try to frame generalizations about functional systems much above the nucleic acids in complexity, the result is always falsity or vacuity. Consider Mendel's "laws" of segregation and assortment: each attributes a functional property to the Mendelian gene. That is, they claim that all members of the functionally identified class, Mendelian genes, have further functional properties by virtue of being Mendelian genes: they assort independently and segregate from homologous alleles. Given that both the antecedent and the consequent classes of these generalizations are identified functionally and therefore their extensions include structurally quite diverse, physically different systems, it would be surprising indeed if these generalizations were laws. For it is highly unlikely that any pair of physical systems identical for any one function should be identical for another. Moreover, if the systems are merely quite similar in fulfilling one function, instead of identical, the chances that they will be even roughly similar in fulfilling another are even lower. So at every level of physical organization where organization reflects numerous selective ties, there will be no very lawful regularities, and such general facts as approximately obtain will not be explainable by any small set of laws about selectively or structurally simpler subsystems. Here we have not only a biological explanation for the impossibility of reduction but an even more fundamental explanation for why there are no strict laws in biology of the sort we are familiar with in physics or chemistry. (And for that matter, we have an explanation for the absence of strict laws in the social and behavioral sciences as well, as we shall see in chapters 7 and 8.)

It is worth remarking here that a rejoinder to this conclusion that advocates the existence of nonstrict laws in biology (and the social and behavioral sciences) is part of the problem with which we are dealing and not an alternative solution. It is tempting to claim lawlikeness for generalizations with disjunctive conditions, ceteris paribus clauses that we cannot finitely discharge and that cannot therefore support the sort of counterfactual claims we make when we don't know which of the disjuncts is realized, or whether ceteris are paribus. But the proponent of the unity of science cannot accept such weak generalizations as *non-strict* laws on pain of surrendering the project of unifying science. Not only can these "laws" not be absorbed by reduction to strict laws, but their weakness makes a permanent barrier to the sort of predictive improvement characteristic of scientific knowledge that the thesis of methodological unity requires.

The alternative, embraced by many philosophers, biologists, and others, is simply to waive the epistemic requirement on predictive improvement and insist that these exception-ridden generalizations are laws just because of the explanatory role they play in biology and elsewhere. Again, John Dupré provides a nice example of this waiver, often based on a tendentious rewriting of the history of science: "Prediction, though long conceived as a very central excellence of scientific understanding, is a goal that has tended to recede rather than approach as various scientific disciplines have increased their understanding of the complexity of phenomena within their domains."[7] First of all, this claim is certainly false for physics and chemistry. Their predictive accuracy has persistently increased in range and depth over the last four hundred years and has increased even more rapidly during the period since the advent of indeterministic quantum mechanics. Second, though some physical phenomena remain as recalcitrant as ever to predictive improvement,[8] among physical scientists the sine qua non of improvement in scientific knowledge has remained improvement in predictive accuracy.[9] Third, as the exponent of the unity of science will insist on

---

7. *Disorder of Things*, p. 3.

8. From Newton's day to our own, physics has been unable to predict the behavior of three bodies acting under mutual gravitational influence, because of the calculational obstacles. The three-body problem is thus an instance of limits on knowledge that reflect computational powers. In recent years, however, chaos theorists have shown that deterministic systems of certain kinds are recalcitrant to predictive improvement, even when we know their fundamental laws of working. But these problems have not dislodged predictive success from its place as the virtue most epistemically prized among physical scientists. In applying lessons from chaos theory to biology, it should be remembered that the theory presupposes that the phenomena do obey strict deterministic laws and that we know what they are. Again, as a result of interaction effects of the variables governing the chaotic processes, predictions are not forthcoming about the process at the level of description that interests us. Should the limitations on our predictive powers in respect of biological processes ultimately be shown to hinge on chaotic considerations, the unity-of-science thesis would, ironically enough, thereby be vindicated.

9. See, for example, Richard Feynman, *QED* (Princeton: Princeton University Press, 1986), chapter 1.

asking, if predictive improvement is surrendered as a mark of improvement in understanding, what mark that distinguishes real improvement from merely apparent improvement in understanding is to be substituted? An appeal to methodological pluralism here is insufficient.

Of course, if biology is one of those sciences that have increased their understanding of the complexity of nature as their predictive powers have receded, or at least in the absence of any progress, then the argument against the unity of science will be powerful even in the absence of an alternative epistemology. But such a claim not only rests on the quite false assumption that biology's predictive powers have not waxed, but also begs the very question against the unity-of-science thesis. After all, if all we can hope for in biology are nonstrict "laws" that cannot be unified with those of physical science, then the burden of proof for the claim that they are laws rests with the opponent of the unity thesis. What is needed is an explanation of why science at the biological level becomes disunified and things become disorderly. Far preferable to seek an account that preserves both unity and order.

This is where instrumentalism comes in. If there are no exceptionless laws and general theories to be found above the macromolecular level, then the generalizations we frame and the theories we employ can reflect only approximations useful for various purposes, models to be applied in some case but not in others, heuristic devices whose limits are known to us through experience but which we can correct and improve in generality only at the expense of their usefulness to us. Thus our aims in biological theorizing cannot include, as those of physical science do, the identification of natural laws of successive generality, precision, and power.

To be sure, uncovering such laws and thereby limning the fundamental nature of reality are not the exclusive aims of the physical scientist. In addition, the physical scientist aims to employ these laws, and for that matter the models, approximations, and simplifications that facilitate the discovery of strict laws, in illuminating and controlling physical processes in harnessing them for technological advance, and in explaining to one another and to nonspecialists the nature of physical processes. In doing many of these things, neither the ultimate laws governing the simple basic constituents, nor the less fundamental laws that can be derived from them, may be necessary or even appropriate. For example, under most circumstances it would be a mistake to explain why a square wooden peg does not fit in a round hole in another piece of wood by appeal to fundamental laws of quantum mechanics, despite the ultimate dependence of the rigidity of macroscopic objects on quantum forces.[10]

Establishing the unity of science is pretty irrelevant to many of these scientific interests, for they have little to do directly with strict laws and derivations from them. Indirectly and eventually, fundamental theory is indispensable, not only to back up the explanations offered in varying contexts, but also to expand our understanding of new phenomena, and finally to tell the truth, the whole truth, and nothing but the truth about nature (or on an instrumentalist interpretation, to give the set of general statements that best combines simplicity and strength in the overall systematization of all observations).

The immediate relevance of the unity-of-science thesis to many of the explanatory tasks of the biologist must also be doubtful, for the same reasons that come into play in physical theory. But it cannot be permanently irrelevant if biology is to be charged, like the physical sciences, with limning the nature of reality. The absence of laws or their irreducibility to those of physical science will either call into question the unity of science (as writers like Dupré imagine) or demand an explanation consistent with it. The instrumentalist thesis that I defend does the latter and so absolves biology of the requirement to search for the truth, the whole truth, and nothing but the truth in the form of laws derivable from those that do reflect the basic character of nature.

On this view the aim of biology must be the sharpening of tools for interacting with the biosphere, but not telling the whole truth about it in a way that can be understood by us or systematized by our fundamental theories. If reductionism is wrong, instrumentalism is right. This, I shall argue, is the consequence of nature's selecting for function and blindness to structure.

---

10. The example is Putnam's. He seems to think that such an explanation would never be appropriate, which surely is an exaggeration. See Hilary Putnam, "Philosophy and Our Mental Life," in *Mind, Language, and Reality* (Cambridge: Cambridge University Press, 1975), pp. 291–303.

CHAPTER THREE

Reductionism and Explanation in Molecular Biology

FROM THE TIME the philosophy of biology absorbed Hull's insight[1] about the impossibility of reducing Mendelian genetics or any of its recognizable successors to molecular biology, it has floundered in its attempt to expound the relation between these two subdisciplines. Some philosophers have worked hard to show how biology redefined the term 'gene'.[2] Others have attempted to weaken the unity-of-science thesis to accommodate the irreducibility of theories in

---

1. *Philosophy of Biological Science*, chapter 1.
2. Perhaps the most promising of such attempts is Waters, "Genes Made Molecular," which makes the notion of "gene differences" central to both subdisciplines. I believe Waters is correct when he argues that "the operative phenotype/genotype relation in classical explanations of inheritance is not a gene/character one (not even a many gene/many character one), but rather a gene difference/character difference relation that is relative both to genetic background in the relevant population and to environmental context." Waters goes on to argue that the molecular gene is a "continuous and interrupted sequence of nucleotide that code[s] for [a] linear sequence in [gene] products," including proteins, messenger RNA (with and without introns, posttranscriptional modified or not), and other posttranslational products. Finally, the gene difference/character difference of classical genetics "is explained at the molecular level: differences in nucleotide sequences of genes result in different sequences in products, or in the case of null mutations, the non-synthesis of products. The resulting differences in products explain the phenotypic differences observed by Morgan and his associates." For reasons that will become apparent, the problem of systematizing these two subdisciplines is not solved but highlighted in Waters' advances on our understanding of how molecular biologists define the concept of gene and his insight about the centrality of *difference* in (observable) phenotypic effects as opposed simply to phenotypes. Phenotypic character differences are, as Waters notes, relative to population and environment, and the blank in 'gene for
     ' can be filled in so many different ways that the only interesting generalizations about genes and characters, or even genes and cytology or physiology, are exception-ridden and unimprovable, of interest only to us, as this chapter argues. For a discussion of how problems of population and environment bedevil our search for biological generalizations, see chapter 5, on the defense of genic selectionism by Waters, Sterelny, and Kitcher.

these two subdisciplines.[3] Still others have continued to seek a sort of reducibility that the relation between Mendelian and molecular biology can realize.[4] It is to the philosophical credit of empiricist philosophers that they have attempted to reconcile these two subdisciplines with the thesis of the unity of science. But the unstable character of results in this debate reflects the failure of empiricist philosophers and biologists to recognize the instrumental character of theorizing in both molecular and population genetics.

This forced choice between reductionism and instrumentalism is perhaps clearest in the most searching of recent treatments of the relation between Mendelian and molecular genetics, Philip Kitcher's "1953 and All That: A Tale of Two Sciences."[5] The view expounded is Hull's original analysis refined by a generation's further work in biology and its philosophy. Kitcher does not argue that his account encourages an instrumentalist view of biological theory. Indeed, at points in his paper, he explicitly rejects it.[6] His arguments against it are unsatisfactory. If I am right, then to the degree Kitcher's analysis reflects the best antireductionist treatment of biological theory, it is after all an argument for instrumentalism.

REDUCTIONISM VERSUS INSTRUMENTALISM

Kitcher's aim is to "account for the almost universal idea that molecular biology has done something important for classical genetics" in spite

3. See Harold Kincaid, "Molecular Biology and the Unity of Science," manuscript, 1991.

4. See, for example, Joseph Robinson, "Reduction, Explanation, and the Quest for Biological Knowledge," *Philosophy of Science* 53 (1986): 333–353; C. Kenneth Waters, "Why the Anti-reductionist Consensus Won't Survive," in *PSA 1990* (East Lansing, MI: Philosophy of Science Association, 1990), pp. 125–139.

5. *Philosophical Review* 93 (1984): 335–373. Pages in text refer to this paper.

6. Elsewhere Kitcher advances views that seem inexorably to lead to instrumentalism or antirealism. In "Explanatory Unification and the Causal Structure of the World" (in Philip Kitcher and Wesley Salmon, eds., *Scientific Explanation: Minnesota Studies in the Philosophy of Science* [Minneapolis: University of Minnesota Press, 1989], 13:410–505), Kitcher argues "that the 'because' of causation is always derivative from the 'because' of explanation. In learning to talk about causes or counterfactuals we are absorbing earlier generations' views of the structure of nature, where those views arise from their attempts to achieve a unified account of the phenomena" (p. 477). On this view it is not the causal order of events in the world that fixes the order of explanation, but the epistemic aim of unification of phenomena under successively simpler theories. In this respect, of course, Kitcher's instrumentalism is not at variance with the thesis of the unity of science.

of the fact that the absence of transmission laws, molecular-Mendelian bridge principles, or derivational explanations makes reductionism (and not just the logical positivist account of it) impossible. What is needed, on Kitcher's view, is a whole new appreciation of what theory is in biological science (and perhaps elsewhere as well).

To begin with, we cannot view Mendelian, or transmission, or what Kitcher calls "classical" genetics as a body of nomological generalizations. Mendel's laws were known to be false almost from the outset of their rediscovery, and no steps were taken to repair them, presumably because such patched-up generalizations were irrelevant to subsequent research in genetics (p. 342).

Instead of a hypothetico-deductive theory (or for that matter a set of models—an approach to be examined in chapter 6), classical genetic theory is to be viewed as a sequence or "linked chain" of "practices": "There is a common language used to talk about hereditary phenomena, a set of accepted statements in that language . . . , a set of questions taken to be the appropriate questions to ask about hereditary phenomena, and a set of patterns of reasoning which are instantiated in answering some of the accepted questions. The practice of classical genetics at a time is completely specified by identifying each of the components just listed" (p. 342). The most important of these is what Kitcher calls the set of patterns of reasoning or argument. These patterns were originally focused on solving "pedigree problems," roughly those surrounding the distribution of phenotypes in successive generations.

Starting with Mendel, recipes for answering questions about pedigrees assumed the single-locus, two-allele case with complete dominance. This argument pattern was superseded by one that accommodates epistasis but is blind to recombination and linkage. With Morgan, a stipulation was added that probabilities of linkage in transmission are functions of colocation on the same chromosome. The original Mendelian pattern included principles of reasoning we recognize as the "laws" of assortment and segregation, but these arguments were surrendered with Morgan for more complex ones that allow for reasoning about phenomena like linkage and recombination.

In addition to the pedigree problem, over time other soluble problems arose in the history of classical genetics, as a result of cytological location of the gene. The theory of gene mapping and of mutations constituted new problem-solving patterns that appealed to and extended the original patterns of the theory of gene transmission.

To illustrate the sort of argument patterns Kitcher identifies as theory, I give an example of the reasoning pattern of classical genetic theory (before the incorporation of linkage and recombination) that Kitcher develops in a subsequent work.

1. There are n pertinent loci $L_1, \ldots, L_n$. At locus L_i there are m_i alleles, $a_{i1}, \ldots, a_{imi}$.

2. Individuals who are $a_{11}, a_{12}, a_{21}, a_{21}, \ldots, a_{n1}, a_{n1}$ have trait P_1; individuals who are $a_{11}, a_{12}, a_{21}, a_{21}, \ldots, a_{n1}, a_{n1}$ have trait P_2; ... {Continue through all possible combinations}.

3. The genotypes of the individuals in the pedigree are as follows: i_1 is G_1, i_2 is $G_2, \ldots, i_N$ is G_N {appended to 3 is a demonstration that 2 and 3 are consistent with the phenotypic ascriptions given in the pedigree}.

4. For any individual x and for any alleles y, z, if x has yz then the probability that a particular one of x's offspring will have y is $1/2$.

5. The transmission of genes at different loci is probabilistically independent.

6. The expected distribution of progeny genotypes in a cross between i_j and i_k is D; the expected distribution of progeny genotypes in a cross ... {continue for all pairs in the pedigree for which the crosses occur}.

7. The expected distribution of progeny phenotypes in a cross between i_j and i_k is E; the expected distribution of progeny phenotypes in a cross ... {continue for all pairs in the pedigree for which crosses occur}. Instructions for filling in the variables above: $L_1, \ldots, L_n$ are to be replaced with names of loci, a_{11}, a_{12}, a_{21}, etc. are to be replaced with names of alleles, $P_1, P_2, \ldots,$ P_N are to be replaced with names of phenotypic traits, $i_1, i_2, \ldots,$ i_N are to be replaced with names of individuals in the pedigree, $G_1, G_2, \ldots, G_N$ are to be replaced with names of allelic combinations (e.g., $a_{11}, a_{12}, a_{21}, a_{22}, \ldots, a_{n1}, a_{n1}$), D is to be replaced with an explicit characterization of a function that assigns relative frequencies to genotypes (allelic combinations), and E is to be replaced with an explicit characterization of a function that assigns relative frequencies to phenotypes.

   Classification: 1–5 are premises; 6 is derived from 3, 4, and 5 using principles of probability; 7 is derived from 2 and 6.[7]

7. Slightly adapted from ibid., 13:439–441.

The work of Morgan substitutes for 5 above the following addition, which incorporates linkage and recombination.

5'. The linkage relations among the loci are given by the equation Prob(L_i, L_j) = p_{ij}. Prob(L_i, L_j) is the probability that the alleles at L_i, L_j on the same chromosome will be transmitted together (if L_i, L_j are loci on the same chromosome pair) and is the probability that arbitrarily selected alleles at L_i, L_j will be transmitted together (otherwise). If L_i, L_j are loci on the same chromosome pair, then $0.5 \leq p_{ij} \leq 1$. If L_i, L_j are on chromosome pairs, then p_{ij} is 0.5. ("Unification," 440)

The reasoning pattern of 5' is refined further to allow for nondisjunction, duplication, unequal crossing over, segregation distortion, and cytoplasmic inheritance. The reasoning pattern of 4 must eventually be surrendered to accommodate the phenomenon of meiotic drive.

What does molecular genetics do for the connected set of reasoning patterns that constitute classical genetic theory? Classical genetics makes certain presuppositions. (Following Kitcher's terminology, a theory presupposes a proposition *p*, if every instantiation of some problem-solving pattern of the theory implies [p. 361].) Kitcher tells us these presuppositions were problematic (classical geneticists had no idea how they could be true): among them is the claim that genes replicate, that mutant genes are viable. These presuppositions seemed impossible, Kitcher alleges, given premises that classical geneticists accepted (p. 361). Molecular biology shows how these presuppositions could be true. It does so by making these conclusions the results of its own patterns of reasoning.

But it does not do so by explaining generalizations of classical genetics, for what molecular genetics explains are not laws in classical genetics. Molecular genetics explains how genes replicate, but "the claim that genes can replicate does not have the status of a law of classical genetic theory. . . . Rather it is a claim that classical genetics took for granted, a claim presupposed by explanations" (p. 361), but not an explicit part of them. Similarly, it explains the problematic presupposition that mutant genes can replicate. It is only these presuppositions of classical genetics that molecular biology explains (by deductive systematization, as reduction requires), but Kitcher argues they are not strictly part of its explanations at all, so evidently molecular genetics explains nothing about classical genetics in its demystification of problematic presuppositions. In the case of mutation, molecular genetics provides a "conceptual refinement" of classical genetics. "Later theories can be said to provide conceptual refinements of earlier theories when the later theory yields a specification of entities that belong to the extension of the predicates in the language of the earlier theory" (p. 364). Molecular genetics provides an account of several different kinds of internal changes that result in mutant alleles, and it enables us to distinguish mutation from recombination. Here again, molecular genetics does something important for classical genetics, but not what reductionism suggests it should do. That mutant alleles can replicate is neither a law of classical genetics nor part of the explanations that its patterns of reasoning provide; it is a problematic presupposition. Or so Kitcher claims.

Only in what Kitcher calls the process of "explanatory extension" does molecular genetics function in something like the reductionist's picture: "[A] theory T' provides an *explanatory extension* of a theory T just in case there is some problem-solving pattern of T one of whose schematic premises [given in T's pattern of argument] can be generated [that is, explained] as the conclusion of a problem-solving pattern of T'. . . . However it does not follow that the explanations provided by the old theory can be improved by replacing the premises in question with the pertinent derivations" (p. 365). Explanatory extension in molecular genetics does not involve general molecular characterizations of all genes, but of particular ones. To cite his example, it enables us to derive the functional differences between sickle-cell hemoglobin cells and normal ones from a specification of the primary sequences of hemoglobin molecules, and the order of basis in normal and sickle-cell hemoglobin genes (together with certain boundary conditions). Explanatory extension in particular cases works because it focuses on differences from other cases, assuming a strong ceteris paribus clause. It does not aim at explaining generalizations about all classical genes, but only particular types, and it can afford to ignore the detailed biosynthetic pathways from genes to proteins, because in these cases the correlation among differences in gene sequences, protein primary structure, and phenotypic effects is relatively straightforward.

Explanatory extension is not reduction, claims Kitcher, for two reasons. First, most classical genetic phenomena can (or can best) be explained only by classical genetic patterns of reasoning, because the phenomena "would look too heterogeneous from a molecular perspective. Intuitively, the cytological pattern makes connections which are lost at the molecular level, and it is thus to be preferred" for explanatory purposes (pp. 371–372). Second, sometimes the direction of explanation in genetics is from classical to molecular, or at least from cytology back to macromolecules: "Understanding the phenotypic manifestation of a gene . . . requires shifting back and forth across levels [of cellular organization]. . . . Sometimes one uses descriptions at higher levels to explain what goes on at a more fundamental level" (p. 371).

Below I adapt Kitcher's account of the reasoning pattern for molecular genetics as it stood in the early sixties.

1. There are n loci $L_1, \ldots, L_n$. At locus L_i there are m_i alleles $a_{i1}, \ldots, a_{im}$.

2. a. The DNA sequence for all is $XYUV \ldots$, the DNA sequence of . . . {continue through all the alleles}.

   b. Details of transcription, post-transcriptional modification, and translation for the alleles in question.

   c. The polypeptides produced by $a_{11}, a_{11}, a_{21}, a_{21}, \ldots, a_{n1}, a_{n1}$ individuals are $M_1, \ldots, M_k$, the polypeptides produced by . . . {continue for all allelic combinations}.

   d. Details of cell biology and embryology for the organisms in question.

   e. Individuals who are $a_{11}, a_{11}, a_{21}, a_{21}, \ldots, a_{n1}, a_{n1}$ have phenotype P_1, individuals who are . . . {continue through all possible combinations}.

3. The genotypes of the individuals in the pedigree are as follows: i_1 is $G_1, \ldots, i_N$ is G_N {appended to 3 is a demonstration that 2e and 3 are consistent with the phenotypic ascriptions given in the pedigree}.

4. If the individual x has $a_{11} a_{12}$ at locus L_1 then the probability that a particular offspring of x will receive all is q_{112}, if an individual x has . . . {continue though all the heterozygous combination}.

5. Same as 5′ above.

6. Same as 6 above.

7. Same as 7 above.

Instructions for filling in schematic letters: Same as above except that X, Y, U, V in 2a are replaced by names of purines and pyrimidines, adenine, cytosine, guanine, thymine, and the M_i in 2c is to be replaced with names of polypeptides.

Classification: 2c follows from 2a and 2b; 2e is derived from 2c and 2d; other derivations are as above. "Unifications," pp. 441–442)

Kitcher notes that the substitution of a variable of probability q_{ijk} for the value of 0.5 in the classical version of 4 above allows molecular genetics to accommodate phenomena like meiotic drive. Molecular genetics is said by Kitcher to provide an explanatory extension of classical genetics' premise 2 into a derived claim 2e. However, reduction is prohibited because "information about cell physiology and embryology [expressed in] (2d) is always too sparse to permit us to make a complete derivation of (2e). One of the closest approximations is furnished by studies of sickle-cell anemia (and of the molecular structure of the genes for globin chains), and such instances show how the derivation would ideally be carried out for phenomena that are developmentally more complex" ("Unification," p. 442).[8]

INFERENCE RULES AND OBJECTIVE EXPLANATIONS

The reader familiar with the history of the philosophy of science will perhaps notice a similarity between Kitcher's picture of the structure of genetic theories and Stephen Toulmin's conception of scientific theories as "inference licenses entitling us to argue from known facts about a situation to the phenomena we may expect in that situation."[9] Following Ryle, Toulmin argued that laws of nature are not true or false, but "inference-tickets," valid in some regions but not in others: "By making the journeys (inferences) so licensed, the physicist finds his way around phenomena."[10]

As a view about the nature of physical laws and an argument for instrumentalism, Toulmin's claims fell out of favor. One powerful argument against them was Nagel's point that the difference between "material rules of inference" and substantive premises is arbitrary, in the context of scientific reasoning.

> [Q]uestions can be raised about a theory when it is regarded as a leading principle that are substantially the same as those which arise when the theory is used as a premise. For whether or not a material leading principle happens to be a theory, the principle is a dependable one only if the conclusions inferred from true prem-

8. For a detailed discussion of both of these examples and their ramifications for reductionism, see Rosenberg, *Structure of Biological Science*, chapter 4.

9. *The Philosophy of Science* (London: Hutchinson University Library, 1953), p. 102.

10. Ibid., p. 104.

ises in accordance with the principle are in agreement with the facts of observation to some stipulated degree. In consequence, there is on the whole only a verbal difference between asking whether a theory is satisfactory (as a technique of inference) and asking whether a theory is true.[11]

If this view is correct, then there is an inevitable tension between Kitcher's claim that there are no lawlike general statements in classical genetics (p. 340) and the claim that there are accepted patterns of reasoning in it.

What seems more reasonable is to hold that the laws about transmission, mapping, and mutation are too complex, too disjunctive, and too numerous for us to discover, express, and employ in arguments more precise than those we actually advance. Thus the patterns of argument we employ are the best we can do, given our cognitive and computational powers. The lawlike statements into which we can "convert" patterns of argument are known to have exceptions, undischarged ceteris paribus clauses, limited ranges from application, and so forth. They are not laws, but they are the best we can do, and given our actual and foreseeable needs and interests, they are sufficiently good. Reductionism thus fails, because there are no *expressible* laws of classical genetic theory to reduce, and the patterns of reasoning that classical genetics employs makes molecular details "irrelevant."

This suggestion gives the reductionist a rejoinder that Kitcher recognizes and attempts to forestall: the claim that molecular genetics is explanatorily irrelevant to the argument patterns characterizing classical genetics will be held by reductionists to "presuppose far too subjective a view of scientific explanation. After all, even if *we* become lost in the molecular details, beings who are cognitively more powerful than we could surely recognize the explanatory force of the envisaged molecular derivation" (p. 348). There are classical laws, albeit complex ones, and we just aren't smart enough to discover them or make use of them. Kitcher responds that this claim "misses a crucial point. The molecular derivation forfeits something important" (pp. 348–349). Roughly, it is blind to important natural kinds of classical genetics, in particular, the class of gene-pair-separation processes (which Kitcher calls PS processes). Classical genetics appeals to cytology to describe

11. *The Structure of Science* (New York: Harcourt, Brace, World, 1961; reprinted, Indianapolis: Hackett, 1983), p. 340.

and explain transmission as a PS process—transmission is explained by bringing it under the mechanism of PS processes. A reduction of classical genetic theory to molecular genetics must preserve the explanatory power of the former theory, and to do this it must preserve the natural kinds of the theory (p. 349).

> However, PS processes are heterogeneous from the molecular point of view. . . . PS processes are realized in a motley of molecular ways.
>
> We thus obtain a reply to the reductionist's charge that we reject the explanatory power of the molecular derivation simply because we anticipate our brains will prove too feeble to cope with its complexities. The molecular account *objectively* fails to explain because it cannot bring out that feature of the situation which is highlighted in the cytological story. (p. 350)

This argument hinges on two crucial unargued assumptions: first, that PS processes are natural kinds and, second, that explanatory success or failure is 'objective'. Both terms, 'natural kind' and 'objective', are undefined in Kitcher's argument, and at least some widely accepted accounts of the meaning of these terms undercut the claims Kitcher makes employing them.

Since Mill's *System of Logic* (1849), 'natural kinds' have been explicated by appeal to laws. The simplest view is that a kind is natural, as opposed to artificial, only if one or more of the properties that characterize all its members figure in a small number of simple general laws. Dupré's statement of the thesis is apposite: what he calls a "strong natural kind" is one "whose essence . . . determines just those properties and dispositions for which it is a matter of natural law that members of the kind will exhibit those properties or dispositions. Thus the real essence of a kind will nomologically determine some range of further properties of its members."[12]

But on this criterion, or on any obvious complication of it, PS processes cannot be 'natural kinds' because, according to Kitcher, there are no laws in classical genetic theory. (Are there laws about PS processes beyond genetics?) If nomological involvement is our criterion of natural kindhood, PS processes don't constitute such a kind, and nomological explanation of instances of PS processes need not preserve a role for the concept of a PS process. Kitcher holds that explanations

12. *Disorder of Things*, p. 65.

must objectively preserve reference to such processes. This suggests that his account of a natural kind ties it more closely to successful or acceptable or 'objective' explanations than to laws of nature. If a kind is natural just in case adequate explanation subsumes diverse explananda phenomena under the kind, then Kitcher's claim will be vindicated: assume classical genetic explanations are 'objective'. Then any other 'objective' explanation of phenomena in the extension of 'PS process' will have to advert to such processes, whether they are nomologically grounded or not. Since 'PS process' is not a natural kind in respect of molecular genetics, it follows that no molecular genetic explanation of phenomena in its extension will be 'objective'. QED.

What could it mean to attribute 'objectivity' to explanations? One sense in which explanations are identified as objective is due to Carl Hempel. In "Aspects of Scientific Explanation" he contrasts pragmatic and objective conceptions of explanation.[13] A pragmatic conception is at least a three-term relation, among explanans, explanandum, and persons. By contrast Hempel seeks to propound an account of explanation that he describes as "objective" in the sense that it "does not require relativization with respect to questioning individuals anymore than does the concept of mathematical proof. It is this non-pragmatic conception of explanation which the covering-law models are meant to explicate" (p. 426). Deductive nomological explanations "are *objective* in the sense that their empirical implications and their evidential support are independent of what particular individuals happen . . . to apply them, and the explanations . . . based upon such laws and theories are meant to be *objective* in an analogous sense" (emphasis added). On this notion of explanatory objectivity, of course the analysis of natural kinds as essential parts of objective explanations reduces to the treatment of them as *nomological* kinds. Accordingly, Kitcher can hardly embrace it.

Nevertheless, Kitcher *seems* to embrace the same sense of explanatory 'objectivity' as Hempel, for he describes as 'subjective' those explanations that work because they are tailored to beings of a limited cognitive power (p. 349); he rejects the notion that molecular genetics fails to explain just because "our brains will prove too feeble to cope with its complexities" (p. 350); he asserts that "the commitment to several explanatory levels does not simply reflect our cognitive limitations" (p. 373). And most important, classical "pattern[s] of reasoning are

---

13. In *Aspects of Scientific Explanation*, p. 426.

objectively to be preferred to the molecular pattern" in explanations of transmission (p. 371, emphasis added).[14]

While Kitcher seems to take seriously the importance of explanatory objectivity in this sense, his arguments against the adequacy of reductionist explanations repeatedly appeal to pragmatic, inquirer-relative, 'subjective' considerations that *do* reflect our cognitive limitations. If the explanatory power of biological theory is ultimately 'pragmatic', however, then Kitcher must in the end tie *biological natural kinds* to these same *cognitive limitations*.[15] It is in this respect that, if correct, the antireductionistic picture of biological science is ultimately instrumentalist.

Kitcher writes that "[c]lassical patterns of reasoning [are] to be objectively preferred to . . . molecular patterns . . . [in the explanation of gamete distribution] because [they] can be applied across a range of cases which would *look heterogeneous* from a molecular perspective" (p. 370, emphasis added). This, as Hull first showed us, is true. But to whom will these cases *look* heterogeneous from a molecular perspective? The answer seems to be, to cognitive agents with our *interests* and our powers. And how much heterogeneity is an obstacle to explanatory success? Well, it depends . . . on the cognitive powers of the agents offering and accepting explanations. After all, agents of our powers can tolerate a certain amount of heterogeneity; those who can keep more alternatives in mind and follow their implications more unerringly can tolerate more heterogeneity in the explanation of gamete distribution and other biological phenomena. Among such cognitive agents

14. It is worth noting that Kitcher attempts to delineate the same nonpragmatic, noncontextual notion of "ideal explanation" in "Explanatory Unification and the Causal Structure of the World": "[O]ne conception of the central problem of explanation—I shall call it the *Hempelian conception*—is the question of defining the class of genuine [explanatory] relevance relations that occur in the ideal why-questions of each and every science at each and every time. We can then suppose that variations in why-questions arise partly from differing beliefs about which topics are appropriate, partly from differing views about the character of answers to underlying ideal why-questions, and partly from differing ideas about that would yield information about those answers" (p. 417).

15. Interestingly enough, this is the approach to natural kinds that Dupré embraces: "What is the natural kind to which [an object] belongs? I claim . . . that such a question can be answered only in relation to some specification of the goal underlying the intent to classify the object" (*Disorder of Things*, p. 5). Dupré does not recognize that thus relativizing kinds undermines his thesis that things in the world are ultimately disorderly, and that therefore our understanding of it can reflect no fundamental unity. If kinds are all and only relative to interests, then it is the heterogeneity of interests that blocks unification, and not the world. Dupré would do well to seek another basis for his denial that kinds are delineated by laws.

many classical patterns of reasoning might well be replaced by molecular patterns. The former patterns won't be reduced to them because they don't reflect laws. Among cognitive agents like us who cannot effectively deal with this much heterogeneity, neither reduction nor replacement is more than an abstract possibility. Instead we seek patterns of reason of the sort Kitcher has identified.

Kitcher holds that, "[i]ntuitively, the cytological pattern makes connections which are *lost* at the molecular level, and it is thus to be *preferred*." Can 'lost' here mean anything other than "lost on us" or cognitive agents no more powerful than we are? I do not think so. That connections at the cytological level are lost at the molecular level cannot mean that there are cytological sequences that are causally independent of molecular processes. As Kitcher notes, "[R]eductionists and anti-reductionists agree in a certain minimal physicalism. . . . There are no major figures in contemporary biology who dispute the claim that each biological event, state or process is a complex physical event, state or process. The most intricate part of ontogeny or phylogeny involves countless changes of physical state" (p. 369). Surely, agents who can tolerate vastly more heterogeneity in explanation than we can could discover and classify together molecular processes that have common effects in cytological connections. On them, such connections would not be lost at the molecular level. For us, however, as Kitcher says, the cytological pattern of reasoning is to be *preferred*. But preference as a criterion of explanatory adequacy can hardly be 'objective'.

Philip Gaspar has argued that there is a sense in which connections made at the cytological level are lost "objectively," at the molecular level, and not just "lost on us."[16] Gaspar begins by claiming that, just as a cytological fact may supervene on any number of different molecular facts, similarly, "the same molecular structure may underlie a multiplicity of higher level" cytological properties (p. 655). For example, a chromosome may give rise to certain effects in a particular set of circumstances, while the chromosomal characteristics supervene on a particular molecular base. But, Gaspar contends,

> citing this base will not adequately capture the relevant [chromosomal] characteristics since these [chromosomal characteristics] will not be the only [cytological] properties that supervene on it [the molecular base.]

16. "Reductionism and Instrumentalism in Genetics," *Philosophy of Science* 59 (1992): 655–670.

Suppose that the cytological properties $P_1, P_2, \ldots, P_k, \ldots P_n$ all supervene on the same molecular base and that some cytological explanation cites one of these properties P_k as causally relevant to the production of some effect E.

Since P_k supervenes on the disjunction of molecular properties $M_1 \lor M_2 \lor M_3 \lor \ldots \lor M_n$, according to Gaspar we must move from the claim that P_k causes E to the claim that the disjunction of $M_1 \lor M_2 \lor M_3 \lor \ldots \lor M_n$ causes E. "The second claim is weaker than the first, and it seems reasonable enough to describe this by saying that the connection made at the cytological level has become lost at the molecular level. Clearly 'lost' here does not mean simply 'lost for inquirers with certain cognitive limitations.' The second claim carries less information than the first for all inquirers, no matter what their cognitive powers" (p. 669). This argument for the "objective" loss of information suffers from a number of problems. Let us distinguish the task of explaining the occurrence of a type of chromosomal event from that of explaining the occurrence of a token of that type. Consider type explanations: if the chromosomal causes supervene on more than one type of molecular state, it is, to say the least, unclear that substituting a disjunction of alternative supervenience bases for a supervening characteristic in an explanation of a general phenomenon constitutes a lessening of information in the original explanation. In fact, molecular biologists strive for such informational accretions.

On the one hand, suppose the explanation offered is of a singular chromosomal effect occurring in a particular cellular system under observation at a particular time—a particular event token, in other words. In place of a singular cytological description of the sole cause, substituting a molecular description in terms of a disjunction of supervenience bases, any one of which were it to obtain would have brought about the effect, might be said to reduce the referential (as opposed to the attributive) information of the original explanation. Presumably, only one of these supervenience bases actually obtained and produced the effect to be explained. On the other hand, it is not clear that such a substitution effects an objective "informational loss." For referential information is of limited import, and explanations of particular event tokens play little role in theoretical science. The referential function of a description is of interest in diagnostic medicine. In diagnostic medicine we do not seek the disjunction of alternative molecular supervenience bases, but the particular disjunct that obtains. In providing it, there is presumably no explanatory loss, but rather an informational gain.

There is a second problem with Gaspar's attempt to show objective informational loss in the transition from the cytological to the molecular. His example trades on equivocations between causally sufficient and causally necessary conditions that vitiate his argument. As Gaspar notes, a chromosome's character leads it to have certain effects "in a particular set of circumstances"; the circumstances and the characteristic are presumably jointly sufficient and individually necessary for the effect they jointly explain. But then the chromosomal character plus those circumstances will not be supervenient on a disjunction of molecular characteristics of genes, but only on one or a small number of them, plus the "particular set of circumstances" molecularly characterized. If the particular set of circumstances is material to narrowing the range of chromosomal effects to the one that actually obtains, then the molecular explanation of this effect must include and specify this same set of particular circumstances in molecular terms. These necessary conditions for the effect must be included in the molecular base on which the chromosomal characteristics that give rise to the effect supervene. Only if the chromosomal characteristic is itself sufficient for its effects, and supervenient on a disjunction of molecular states, will there be the sort of informational loss Gaspar contemplates. However, if, per impossible, a particular chromosomal characteristic in a particular set of circumstances invariably gives rise to a certain effect, then an information-preserving reduction should be in the cards.

Finally, the real trouble with Gaspar's example is that, as Kitcher notes ("Unification," p. 442), there are no invariable relations at the cytological level on which to ground the original cytological explanation—there are no cytological laws, and no biconditionals between cytology and molecular biology, and so no basis to suggest that there is objective informational loss between generalizations at the cytological level and those at the molecular level. I conclude that any such informational loss is loss "for us"; we surrender descriptions convenient for organizing our experience and predicting at levels of accuracy good enough for our concerns, for descriptions that figure in generalizations and theories too complicated for us to employ in real time to meet our practical needs to organize our experience.

From the claim that cytological connections are lost at the molecular level, Kitcher infers that "explanatory patterns that deploy the concepts of cytology will endure in *our* science because we would forswear significant explanatory unification . . . by attempting to derive the conclusions to which they are applied using the vocabulary and reasoning patterns of molecular biology." This seems to be correct, but only on the understanding that *our* science in part reflects our cognitive limitations. Yet this is not Kitcher's conclusion. Instead he draws the moral that "the current divisions of biology [are] not simply . . . a temporary feature of our science stemming from our cognitive imperfections, but . . . the reflection of levels of organization in nature" (pp. 370–371).

CONCLUSION

The complexity that selection for effects produces above the level of the molecule makes Kitcher's account of the structure of classical genetics and its relation to molecular genetics a compelling one. It explains why we can't discover classical laws and must resign ourselves to patterns of argument. If we accept Kitcher's treatment of classical genetics as characterized not by laws but by principles of argument, and take seriously his denial that there are laws in classical genetics, then despite his claims to the contrary, we are committed to treating biology as an *instrumental* science, that is, as a body of claims each of which is qualified by an implicit appeal to its usefulness for cognitive agents of our powers. For each material argument pattern in classical genetics there is a substantive statement that, though it is no law or even an approximation to one, is the most useful hypothesis for dealing with some problems of interest at some level of the development of our knowledge. Reduction is impossible because the generalizations about classical genetics supposed to be reduced are unavailable. This impossibility is no defect in classical genetics, however, for that theory is not a body of laws but primarily a set of patterns of argument, patterns warranted by their usefulness in particular problems and at particular stages of biological development.

The reductionist's rejoinder, that there must in principle be such laws, though they will be too disjunctive, complex, and heterogeneous to be of much use to *us,* is to be met with a shrug of the shoulders. Yes, what of it? Admitting this is of no moment to the methodology of biology; it has no consequences, practical or theoretical (though it might have some metaphysical consolation).

Thus it turns out that not only the content but also the character of biological theorizing is contingent on biological facts. Biology is at its best an instrumental science because of the operation of biological forces. First, the complexity of nature above the level of the molecule is the result of selection for function and its blindness to structure. Second, the biological fact is that the sentient creatures who develop the subject have (through selection) come to have cognitive powers that limit their ability to deal systematically with this amount of complexity.

Now we understand why the smooth reduction of biological theory to physical theory is not on the cards. Our understanding is compatible with, indeed rests on, a materialist and mereological determinist approach to biological systems: they are, as we thought, "nothing but" physical ones, even though we cannot systematically derive the biological from the physical. We also understand now why the doctrine of the unity science must be qualified. Its epistemological requirements need no qualification. Biology fully honors the requirement of empirical evidence. But its demand that we systematize theory needs to be qualified, and the goals set for the unity of science need to be restricted above the level of physical theory.

We cannot expect ever more accurate biological explanations of biological phenomena, or more accurate physical explanation of biological regularities. Above the level of the physical sciences, the only sciences we can contemplate are strongly contingent on our cognitive and computational powers. Such generalizations, though useful for us, are not subject to corrections we can provide that will enable them to be unified with the rest of science while retaining their usefulness to us.

Accordingly, in pursuing biological theory we are constantly faced with a choice between usefulness, given our limits and interests, and greater systematization in the direction of the truth, the whole truth, and nothing but the truth about the nature of things. This is not a choice that the unity-of-science doctrine can contemplate with equanimity. It holds that the sciences all share an equal commitment to the goal of a completely accurate account of reality (or at least an empirically adequate one). It allows for models that are heuristically useful, but does not condone the suggestion that in principle the shape of scientific theory should be determined by considerations about our limits, instead of the evidence. On the other hand, proponents of the unity of science can accept such an outcome, in practice. If they can be assured that it is a consequence of contingent facts about the world, as opposed to some philosophical theory that sets limits to the writ of empiricist epistemology and materialist metaphysics, the proponents of the unity of science can accept limitations on our scientific knowledge.

Proponents of the unity of science must accept that biology and the other sciences that rest upon it are, in effect, sciences implicitly about *us*, in addition to being explicitly about nature. Fully to explain theories in these sciences will require an understanding of our cognitive character. This does not make them subjective in the sense of doubtful, though if we did not exist, biological theory would be at best only counterfactually true. It would be a set of statements about what generalizations and theories would be most convenient for agents of our cognitive limits to believe. If there were no such agents, our biological theory would not be a set of generalizations—nomological or otherwise—true about the actual world. In this regard, scientific realism of the sort tenable with respect to physics and chemistry is not tenable with respect to biology. But the thesis of the unity of science as a metaphysical claim applies only to theories about which scientific realism is tenable— theories that claim to describe the way the world is, independent of our thoughts about it. It is perfectly consistent with the most stringent version of a thesis of the unity of science for scientists to have recourse to models, case studies, idealizations, and approximations useful for technological and other applications. Such recourse will characterize many parts of chemistry and physics, and *all* of biology.

Evolution, Drift, and Subjective Probability

EVOLUTIONARY BIOLOGY is a statistical subject par excellence. In fact, many of the most important statistical methods commonplace in the social sciences were invented for evolutionary inquiries by Galton, Pearson, Neyman, Haldane, Fisher, and Wright. But the exact place of stochastic processes in evolution, and of statistical concepts in evolutionary theory, has remained controversial. The controversies in biology have been substantive ones about the relative contributions of selective and sampling effects to diversity over time. In philosophy the controversies have surrounded the kind of probability to which evolutionary theory adverts. In this chapter I endeavor to show that the kind of probability that figures in evolutionary theory does *not* figure in the evolutionary process, and that this makes the theory biologists employ a useful instrument reflecting our cognitive and computational limitations. Once we see this, it becomes misleading to say that the process of evolution is stochastic at all.

STATISTICAL MEASURES AND CENTRAL TENDENCIES

Charles Sanders Peirce first argued that the theory of natural selection is a statistical theory, for reasons much like those that make thermodynamics a statistical theory. In particular, the theory claims that fitness differences are large enough and the life span of species long enough for increases in average fitness always to appear in the *long run*; and this claim, Peirce held, is of the same form as the statistical version of the second law of thermodynamics. The second law also makes a claim about the long run, and its statistical character is due to this claim: thermodynamic systems must in the long run approach an equilibrium level of organization that maximizes entropy. Over finite times, given local boundary conditions, an isolated mechanical system, like the molecules in a container of gas, may sometimes interact so as to move the entropy of the system further from, instead of closer to, the equilibrium level. But given *enough* interacting bodies and *enough* time, the system will always eventually move in the direction prescribed by the law. Thus we can attach much higher probabilities to the prediction that nonequilibrium systems will reflect greater entropy in the future than we can to predictions that they will move in the opposite direction. As we increase the amount of time and the number of bodies interacting, the strength of the probability we can attach to the prediction becomes greater and greater.

The same kind of probabilistic claims that the second law of thermodynamics makes about the direction of entropy change is made by evolutionary theory about the direction of fitness changes. Evolution need not and does not move in a straight line toward equilibrium levels of populations for various species and their subpopulations. The theory asserts that, over the long run, evolution must move in the direction of equilibrium, and that the length of time it takes to get there is a probabilistic function of fitness differences and population sizes.

This account[1] of the statistical character of the theory is at least quite incomplete. It is certainly misleading. To begin with, the analogy to the statistical character of thermodynamics is not very illuminating, because the statistical character of the second law is itself not well understood. Although the law makes a probabilistic claim, it is supposed to reflect fully deterministic behavior among the constituents of thermodynamic systems, in accordance with Newtonian mechanics. Yet the reduction of thermodynamics to mechanics has never been satisfactorily effected in the general case.[2] So there is no general explanation of how thermodynamic probabilities emerge from mechanical certainties. Nor is it clear what kind of probability the theory trades in: does it make a claim about epistemic probabilities, probabilistic propensities, or long-run relative frequencies, or is there some other interpretation of probability that is most suitable to expressing its claims?

In the absence of solutions to these problems, an improved understanding of the statistical nature of evolutionary theory must proceed from a direct examination of the theory and not from an analogy to a

---

1. Advanced in David Hull, *Philosophy of Biological Science,* to which I subscribed pretty completely in *The Structure of Biological Science.*

2. This is a small embarrassment for the thesis of the unity of science. It is small because no one supposes that it warrants the claim that thermodynamic properties of systems are emergent from or otherwise independent of the mechanical properties of their components.

statistical theory that is not itself well understood. This direct examination, however, will eventuate in an account of the role of probability in evolutionary biology that strongly reinforces the conclusion of chapter 3. It will turn out that the only probabilities to which the theory is committed are the subjective probabilities that agents of our cognitive powers require to apply the theory to actual processes among populations of interest to us. If people were a lot smarter, there would be proportionately less reason to appeal to such epistemic probabilities and mutatis mutandis less reason to treat the theory as statistical.

The question of whether evolutionary phenomena are stochastic is different from the question of whether *our best theory* of these phenomena is unavoidably statistical. Our best theory, present or future, may turn out to be statistical because the deterministic facts about evolution are beyond our cognitive and computational powers to apprehend in useful terms. Recent work in chaos theory has revealed a variety of deterministic systems about which our best theories will be unavoidably statistical. Or rather, there will be some deterministic properties of such systems about which our best predictions will be probabilistic. Here the features that lead to such limitations are, as in the three-body problem of mechanics, the infinite iteration of interaction effects that prevent our calculations from coming to a conclusion arbitrarily close to the actual value of the variable we wish to predict. The indeterminacy here is in ourselves and not in the things in the world.

Thus we need to address separately the questions of (1) whether the process of evolution is deterministic or probabilistic and (2) whether our theory of natural selection, as currently understood, is deterministic or probabilistic. Addressing either question requires us to identify the evolutionary features or "state descriptions" with respect to which we make claims of determinism or probabilism. As Nagel pointed out long ago,[3] a theory, for example quantum mechanics, can be deterministic with respect to one set of state descriptions (the quantum mechanical ones), and indeterministic or probabilistic with respect to another set (the classical or Newtonian ones). The psi function, which gives the quantum mechanical state of a system, varies deterministically in accordance with the Schrödinger wave equation. But the position and momentum of the same system, which gives its Newtonian state description, varies only probabilistically in accordance with the absolute value of the square of the psi function.

3. *Structure of Science*, pp. 285ff.

This requirement that we specify state descriptions confronts us with a difficulty, for it is by no means clear what the appropriate state descriptions are, with respect to which we should pose the questions of whether either the process of evolution itself or the theory of natural selection is deterministic or not. The philosophically least controversial assumption about the appropriate state description for our purposes is that the states to be systematized by evolutionary theory are proportions of genetically similar population within and between species, that is, gene frequencies.

Then there is the question of what interpretation to attach to the probability operators that are ubiquitous in evolutionary discussions. Are they epistemic or long-run relative-frequency claims? Or do they reflect some sort of objective propensities? Two sorts of interpretations seem to be ruled out straight away.

First, we can rule out the notion that gene frequencies might be probabilistic for reasons having to do with the indeterministic character of the fundamental physical processes on which evolutionary phenomena supervene. The world is certainly indeterministic in its fundamental laws of working, and since fundamental microphysical processes are hooked up to macroprocesses in ways that convey their microindeterminism to the macroworld (for example, in Geiger counters), phenomena at every level of organization and aggregation are "infected" with quantum mechanical indeterminism. The only sort of probability that seems to many to make sense of quantum mechanical indeterminism is that which is an unanalyzable propensity or a disposition of events to bring about other events. Accordingly, some evolutionary events must have these probabilistic propensities to result in evolutionarily significant consequences, just because they are built up out of or hooked up to quantum events. More concretely, point mutations of the genetic material are well known to be the result of probabilistic quantum interactions. Thus at least some mutationally driven evolution will be probabilistic for the same reason physics is.

This admission settles no problem of interest to evolutionary biologists. Mutation is but one among several sources of evolutionary change and certainly not the most important one. Indeed, evolution can proceed in the absence of mutation and would do so even if mutation were a thoroughly deterministic process. Like other sources of variation, mutation is required by Darwinian theory to be random with respect to the environmentally adaptive needs of organisms, just as the fall of a pair of dice is random with request to the hopes of dice throwers. Sources of variation do not have to be indeterministic any more than roulette wheels have to be radically indeterministic. In general the quantum probabilities involved in biological processes are so small, and the asymptotic approach to determinism of everything physical above the level of the chemical bond is so close, that quantum mechanical probability could never explain the probabilistic character, if any, of either evolutionary phenomena or evolutionary theory. That is, quantum mechanics cannot explain the probabilistic character of evolutionary theory if, as the unity-of-science doctrine holds, the theory describes purely physical processes at the level of aggregation where selection for effects begins to intervene. By the time nature gets to this level, it has long since asymptotically approached determinism.[4]

Second, we can rule out the notion that gene frequencies reflect purely epistemic probabilities. If the *only* probabilities at work in evolutionary theory are epistemic, that is no reason to suppose that evolutionary phenomena are in themselves indeterministic or probabilistic. Epistemic probabilities are "subjective." They are relations between events and ourselves, or more exactly, between events and our beliefs, our evidence. Subjective probabilities measure degrees of belief, the betting odds we are willing to give that reflect how much evidence we have for a certain conclusion, assuming that a payoff of a certain amount of money or some other good hinges on our being right. If there were no epistemic agents, there would be no epistemic probabilities. Since evolution can and does take place in the absence of epistemic agents, the process could not involve any such epistemic probabilities. Another somewhat tendentious way of saying this is that, if evolutionary processes are stochastic, the sort of probability in question must be "real" and "objective," not "subjective."

Of course, even if epistemic probabilities are no part of the process of evolution, they may still figure ineliminably in our best theory of evolution (as they do in theories of certain deterministic but predictively "chaotic" processes). But if epistemic probabilities do figure in our best theories, then as noted above we shall have to rethink seriously the nature of that theory, which will then be in part about our beliefs and their relation to a world that is independent of our beliefs.

---

4. Even Dupré seems to concede this much (*Disorder of Things*, p. 177).

STATISTICAL MEASURES DO NOT
A STATISTICAL THEORY MAKE

One way to argue that the theory of natural selection is a statistical theory is to note its reliance on statistical measures of its data and to show that such measurement is intrinsic to the theory. Kim Sterelny and Philip Kitcher argue that evolutionary theory is a statistical theory at least in part because of its use of statistical fitness coefficients to represent the expected survivorship and reproductive success of organisms. The employment of such statistical variables, they write, reflects a

> strategy of abstracting from the thousand natural shocks that organisms in natural populations are heir to. In principle we could relate the biography of each organism in the population, explaining in full detail how it developed, reproduced, and survived, just as we could track the motion of each molecule of a sample of a gas. But evolutionary theory, like statistical mechanics, has no use for such a fine grain of description: the aim is to make clear the central tendencies in the history of evolving populations, and to this end, the strategy of averaging . . . is entirely appropriate.[5]

Note that this passage attributes aims and strategies to a theory, instead of attributing them to theorists who make use of it. Why then does statistics enter into the theory? Sterelny and Kitcher say it enters because the theory has "no use for" a fine grain of description; the theory's aim "is to make clear the central tendencies in the history of evolving populations" and its strategy is one of "abstracting from the thousand natural shocks that organisms . . . are heir to."

This claim betrays an instrumentalist approach to biological theory, which Kitcher and Sterelny may not intend but which certainly jibes with the instrumentalist interpretation I tried to saddle on Kitcher in chapter 3. Recall the claim that cytological patterns would be "lost" at the molecular level. Sterelny and Kitcher's claim here echoes this thought, and I shall argue it is subject to the same instrumentalist interpretation.

Having no use for fine-grained description but instead an interest in

5. Kim Sterelny and Philip Kitcher, "The Return of the Gene," *Journal of Philosophy* 85 (1988): 345.

central tendencies reflects one of the aims of evolutionary theor*ists*. Theories do not have aims as such, except perhaps those of truth or empirical adequacy, and these aims are not served by neglecting fine-grained descriptions or focusing only on central tendencies. Theorists have aims, which along with theorists' means—their capacities for dealing with data—jointly dictate their strategies. In the case of the statistical character of the theory of natural selection, Sewall Wright effectively made this point long ago.

> Natural selection is an exceedingly complex affair. Selection may occur at various biological levels—between members of the same brood, between individuals of the same local population, between such populations (as through differential increase and migration) and finally between different species. . . . Selection among individuals may relate to . . . mating activities . . . [, to] differences in attainment of maturity, to differential fecundity and to differential mortality. Selection may act steadily or may vary both in intensity and direction. . . .
>
> In such a complex situation, verbal discussion tends towards a championing of one or another factor. We need a means of considering all factors at once in a quantitative fashion. For this we need a common measure for such diverse factors as mutation, crossbreeding, natural selection and isolation. At first sight these seem to be incommensurable but if we fix our attention on their effects on populations, rather than on their own natures, the situation is simplified. Such a measure can be found in the effects on *gene frequencies*.[6]

As the passage makes clear, it is not the aim of the theory of natural selection to explain central tendencies; it is *our* aim, to which we apply the theory. Were our aims different, or our means of attaining these aims different, our employment of the theory might not be statistical or might involve the employment of different statistical measures. Indeed, were our capacities different, our identification of "central tendencies"—statistical or otherwise—would be different too.

Wright's point is that statistical description is introduced so that we can measure the combined effects of diverse evolutionary forces by providing a quantitative description of their *effects* on the distribution

---

6. "Statistical Genetics and Evolution," Gibbs lecture, 1941, reprinted in *Evolution*, ed. William B. Provine (Chicago: University of Chicago Press, 1984), p. 468.

of traits. The theory does not "abstract . . . from the thousand natural shocks that organisms . . . are heir to." *We* abstract from the description of the effect of these shocks on organisms. The theory of evolution is statistical because averages are employed to describe its explananda phenomena. Natural selection in itself need not be an indeterministic process or a stochastic one just because we must rely on statistical measures to identify its effects in large populations. Selection operates on individual items *one at a time*. Thereby it produces effects that can be summed and from which average values, mean values, variances, and other statistical properties of whole populations can be constructed. These statistical properties and changes in them describe the explananda of evolutionary biology; they are causally inert, epiphenomena, which we track because they provide a convenient description of the course of evolution, a description convenient *for us*.

If without these statistics that sum and combine the effects of diverse evolutionary causes *we* would miss interesting evolutionary generalizations, then these are generalizations that are of interest *to us*, because of our cognitive and computational capacities. Cognitive agents with greater powers to "crunch" data, to track the fate of larger numbers of smaller lineages over more fine-grained periods of time, would compute averages and statistical measures that more closely reflected the processes that underlie and explain the "tendencies" that we identify as "central" because of our cognitive limitations. A cognitive agent of unlimited powers, a Laplacian demon, would have no need of statistical information to explain evolutionary processes by appeal to natural selection over blind variation. Or so I shall argue.

Sterelny and Kitcher are correct to say that thermodynamics has no use for fine-grained description, like that involved in tracking the motion of a molecule. But the parallel breaks down just because an evolutionary biologist of even our limited cognitive powers must be interested in the fate of individual organisms, while the thermodynamic theorist cannot have any interest in the individual molecule, for it has no thermodynamic properties. By contrast, every organism has a property that figures in evolutionary theory: fitness. And a single individual variation, like the first white eye Morgan ever noticed on a fruit fly, can amplify over generations until it is the property whose representation in the population becomes the statistical fact to be explained. The properties that do the explaining—fitness and fitness differences—are property of individual items, whose arithmetic average value for the whole population has itself no explanatory role beyond its summing of individual values. By contrast, entropy is a thermodynamic concept that has withstood straightforward reduction to mechanical properties of constituent molecules and their aggregations, from Boltzmann's time to our own. If thermodynamics is statistical only because temperature is an aggregate statistical measure of the effects the theory explains, then evolutionary theory cannot be statistical for the same reasons thermodynamics is statistical.

Could the fitness (or fitness differences) of individual organisms (or other units of selection, if there are any) be stochastic? And could natural selection be stochastic because of it? Could the theory be statistical because of the probabilistic character of fitness? As John Beatty has shown,[7] statistical measures of fitness certainly introduce effects impossible to distinguish from statistical randomness. Consider the measurement of fitness by appeal to probabilistic propensities. If we assign a normal distribution around a single number of offspring as maximally probable, Beatty argues, "it is difficult to distinguish between random drift and the improbable results of natural selection."

> Whenever there are [probabilistic] fitness *distributions* associated with different types of organisms, there will be *ranges* of outcomes of natural selection—some of the outcomes within those ranges will be more probable than others, but all of the outcomes within the ranges are possible outcomes of natural selection. And yet some outcomes within a fitness distribution (the outer-lying outcomes of a bell-shaped fitness distribution, for instance), are in a sense "less representative" of the off-spring-contribution abilities of the organisms in question. (p. 197)

Beatty explains this point with a coin-tossing analogy: it is improbable but possible that flipping a fair coin one hundred times will produce ninety tails and ten heads, even though such an outcome is unrepresentative of the fairness of the coin. Similarly, an organism with fitness level measured by a bell-shaped distribution around, say, five offspring may leave ten even though it is "less representative" of the fitness of the organism.

7. In "Chance and Natural Selection," *Philosophy of Science* 51 (1984): 183–211. This excellent treatment of the subject repays careful study. Pages in text refer to this work.

To the extent those outcomes are *less* representative of the physical abilities of those organisms to survive and reproduce in the environment in question, any evolutionary change will be *more* a matter of random drift. In other words, it seems we must say of some evolutionary changes thay they are to some extent, or in some sense, a matter of natural selection *and* to some extent, or in some sense, a matter of random drift, and [one of the reasons] we must say this is that it is conceptually difficult to distinguish natural selection from random drift, especially where the *improbable results of natural selection* are concerned. (p. 197)

Of random drift much more is to be said immediately below. Meanwhile the crucial point to note is that the probabilistic character of evolutionary changes follows only on the controversial condition that fitness is in fact a probabilistic propensity, as opposed to the less controversial supposition that fitness is merely *measured* by units of such a propensity. If probability provides the measure of fitness, then probabilistic outcomes in evolutionary predictions or explanations will be a consequence of probabilistic measurements of boundary conditions, not of the stochastic character of those conditions. In this case the right conclusion is that the theory is statistical because for us the most convenient measurement of its causal variable is probabilistic, as Sewell Wright suggested. It does not follow that evolution is an irretrievably stochastic phenomenon.[8]

For natural selection to be inevitably probabilistic, selection itself has to be a stochastic matter. For the theory to be statistical, the statistical description of the impact of selection has to be indispensable. And for the statistical character of the theory to reflect the stochastic nature of the process, the theoretical indispensability must match the natural inevitability. The convenience that statistical aggregates provide in measuring the combined causes and effects of selection does not meet this standard for any explanation of the statistical character of the theory of natural selection. But what of that paradigmatically statistical concept, evolutionary drift?

8. I return to the issue of whether fitness is a probabilistic propensity in chapter 6. It is worth noting that Beatty has come to doubt that fitness is such a probabilistic propensity. See John Beatty and Susan Finsen, "Rethinking the Propensity Interpretation," in Michael Ruse, ed., *What the Philosophy of Biology Is* (Boston: Kluwer, 1987), pp. 17–30.

WHAT EXACTLY IS DRIFT?

Evolutionary biologists seem to identify the source of unpredictability in evolution with the phenomenon of drift. For example, Strickberger writes that mutation, selection, and migration

> act in a directional fashion to change gene frequencies progressively from one value to another. . . . whether going towards fixation or equilibrium, the constancy of these forces enables them to be described as *directional*. . . . In addition to these directional forces, however, there are also changes which are not tied to the gene frequencies involved. Because of this, such forces cause gene-frequency changes that can go in one direction or another, from generation to generation, *without any predictable constancy* [emphasis added]. One of the most important of such *nondirectional* forces arises from variable sampling of the gene pool each generation and is known as *random drift*.[9]

According to Strickberger, genetic drift results when the effective population size (roughly the number of parents in a population) is small enough for gene frequencies to vary as a result of "sampling error." In this context "sampling error" has a misleading connotation, since it suggests investigator intervention that might mistakenly select for experimentation or some further treatment a nonrepresentative sample from a population. "Sampling error" is to be understood here in terms no more anthropomorphic than the meaning of "selection" in evolutionary theory. It is "nature" that picks a subset of the population for further treatment; if its selection is not representative of the whole population, the results may be different from what they otherwise would be. And as all experimenters know, the smaller the sample, the higher the probability that it will be unrepresentative.

Only a subset of the members of each generation of a population reproduce. This subset constitutes the "sample." Reproduction constitutes the "treatment." The smaller the sample is, the less likely it is to be representative and the larger the "sampling error." As long as the effective population size is very large, in the absence of selective forces, successive generations will deviate only a little in gene frequency. If standard deviation from previous generations is measured as $\sigma = \sqrt{pq / 2N}$, then if N is large, the frequencies of two alleles, p and q,

---

9. *Genetics*, first edition (New York: Macmillan, 1968), p. 736.

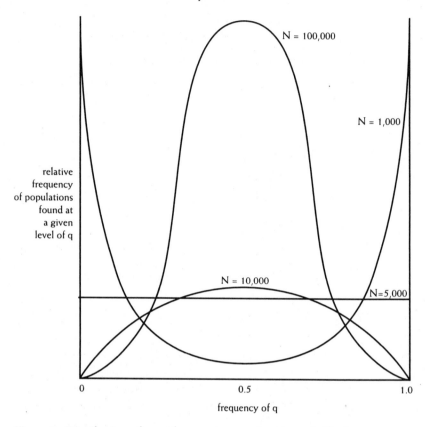

Figure 1. Distribution of gene frequencies as a function of effective population size.

will remain close to their starting frequencies approximately two-thirds of the time (two standard deviations, in statistical parlance). When N is small, the frequencies will fluctuate widely and will frequently go to fixation for one of the two alleles, and extinction for the other. If a species is composed of a large number of small isolated subpopulations, then the result may well be "fixation" for p among half the subpopulations and for q among the other half. Only one of the two genes will be represented in any of the subgroups. This may happen even if the starting frequency of each allele in the whole population is .5. Strickberger reproduces a graph from Sewall Wright (1951) to illustrate the effects of drift for different population sizes (N = 100,000, 10,000, 5,000, 1,000) when selection is zero, migration is small, and the starting frequency for q is .5. In the largest population (N = 100,000) the frequencies of p and q remain close to .5. In the smallest population (N = 1,000) the proportion of populations in which both alleles are equally represented is quite small. It is clear that, even when the total population of a species is very large, the number of conspecifics with which one organism can interact throughout its entire lifetime may be quite small. This possibility provides wide scope for drift. As Strickberger notes, according to Sewall Wright a good deal of change in gene frequencies over time, even in the face of selective forces, may be due entirely to the effects of random drift on a population whose "structure" reflects such isolated subgroups. A hundred populations of 1,000 may be equal in total numbers to a single population of 100,000, but as the graph shows, starting from the same gene frequencies as the large population, they may reflect a quite different set of gene frequencies from that of the large group after a relatively brief period.

Our reliance on gene frequencies in measuring the effects of selection is certainly responsible for the probabilistic character of many of the theory of evolution's characteristic claims. But that is certainly not enough to say the theory is probabilistic. Like probabilistic measurements of the causes of selection in fitness differences, probabilistic measurement of its effects is no reason to suppose that the phenomenon of evolution is itself stochastic. Compare the situation in physics.

The application of Newtonian mechanics to astronomy, or for that matter to any system that requires sampling and so may introduce measurement errors, requires an appeal to statistical considerations and produces results that are probabilistic. Given a probability distribution of the positions and momenta of the bodies in a Newtonian mechanical system, we can predict with a probability as close to one as we like, that probability distributions of position and momentum at any future time will be equally arbitrarily close to those the deterministic equations of Newtonian mechanics lead us to expect when applied to the original probability distribution. Yet these facts have no tendency to show that Newtonian mechanics is probabilistic. Of course if, as it turns out, position and momentum are not just epistemically undetermined but physically undetermined, as quantum mechanics requires, then the most complete theory governing these variables will be indeterministic, for reasons above and beyond our inability to give more than probabilistic estimates of exact position and momentum.

Moreover, even the possibility of small errors of measurement can make a deterministic system utterly unpredictable in its behavior. In this respect, whether a theory is deterministic or not is independent of whether it enables us to make well-confirmed predictions about states of the system it describes. If our measurement of initial conditions must be probabilistic because of limitations on our measuring devices, then the theory's predictions will be probabilistic at best. At worst, as chaos theory reminds us, the theory will make no definite predictions at all, because of magnifying interaction effects between variables we can measure only probabilistically and not accurately enough.

It seems clear that probabilistic measures of gene frequencies are of the measurement-error sort. They reflect the degree of our confidence that we have avoided large "measurement error." So these probabilities cannot be the source of the statistical character of the theory of evolution, still less of the phenomena of it. We measure the proportion of genes in terms of relative frequencies, not because these proportions are indeterministic but because the only feasible means of applying the theory of natural selection is to sample the population and infer the distribution of genes. This involves experimenter or observer sampling error. But there is no reason within the theory of natural selection itself to suppose that the proportion of different genes is undetermined, or inherently probabilistic. Indeed, as I argued in *The Structure of Biological Science*, since the theory of natural selection itself makes no mention of genes but only presupposes some vehicle for hereditary transmission, indeterminism of any sort at the level of genes, including Mendelian segregation and assortment, cannot be responsible for the probabilistic character, if any, of natural selection *in general* or of the theory of natural selection considered in isolation. If probabilistic methods are forced on us *only* because we have to apply the theory of natural selection to events that are probabilistic as a result of Mendelian assortment and segregation, then neither evolution as a general phenomenon nor the theory itself is inherently probabilistic. Only its local applications turn out to be. In a biosphere with different, much simpler hereditary mechanisms, change in trait frequencies could be thoroughly deterministic.

Might random drift figure in our theories in the same way our probabilistic measures of gene frequency do, reflecting our ignorance of initial conditions, and not as an essential probabilistic element in the theory of natural selection? Is drift a matter of epistemic probabilities touching on the initial conditions to which we apply the theory? There are some tempting reasons to think this is the right view. First consider the minor terminological note. Strickberger identifies selection, mutation, and migration as evolutionary forces, which determine the direction of evolution, by contrast with drift, which he declines to identify as another *evolutionary* force. Gene frequencies *change* under the effects of drift, but they may not be said to *evolve*, that is, to show changes of apparently adaptational significance. Indeed, insofar as evolution means adaptational change, drift is clearly no part of it, for no one identifies the source of adaptation in drift alone. Could drift actually be a way of referring to those unknown *nonevolutionary* forces that interfere and deflect evolution from the outcomes that deterministic forces like selection, mutation, and migration would otherwise secure? Alternatively, could drift be a way of dealing with those evolutionary forces that escape our notice because of their irregularity and subtlety? If either of these alternatives is right, then drift provides no reason to think evolution "in the objects" is in fact probabilistic. If the second view is right, then drift would be a source of indeterminism only within the theory, without reflecting any probabilities in the phenomena.

ALTERNATIVE INTERPRETATIONS OF DRIFT

Suppose there is a small population of giraffes, whose longest-necked members are most well adapted to the veldt and are secretly removed by poachers to zoos just because they have the longest necks, there to languish, leaving no offspring. As a result, gene frequencies change in the small population. Short-necked giraffes come to predominate in the group for several generations. What are the game-preserve biologists to say? Is the change in gene frequencies a case of drift? Since the physical environment has not changed and the biologists know nothing of the poachers, their choice seems to be either drift or disconfirmation of the theory of natural selection. Naturally, our biologists choose drift. In so choosing, are they embracing the hypothesis that nonevolutionary forces interfered to move the population away from an adaptational equilibrium? Or are they adopting the idea that as yet unrecognized environmental factors did so, shifting the locus of the adaptational equilibrium and moving the population toward it?

Well, if the naturalists begin a careful study of the environmental conditions of giraffes, hoping thereby to explain the change in gene frequencies in terms of a change in the adaptive topography, then evidently they adopt the latter idea. If so, either the naturalists do not grant that drift played a role, or else for them drift *means* simply movement of gene frequencies in adaptive but unexpected and temporarily inexplicable directions. If so, drift hypotheses are introduced to explain changes in gene frequencies as expressions of our ignorance of deterministic evolutionary forces, and drift does not confer any real probabilistic features on the phenomena of evolution or on theory of natural selection. It simply reflects our ignorance.

Some commentators would stigmatize this use of drift as hopelessly "Panglossian."[10] These biologists condemn the strategy of always seeking a new adaptational explanation whenever a previous one has been placed in doubt. Treating drift as a cover for unknown adaptational forces, to be cited whenever gene frequencies do not evolve in the evolutionarily expected direction, deprives the theory of much of its empirical content, on these views. Such opponents of "Panglossianism" will certainly seek a notion of genetic drift that does not condemn them to this adaptational imperative. They are committed to saying that drift either is or reflects the operation of other, nonevolutionary forces. Thus Lewontin and Gould say specifically that alleles can become fixed in a population through drift "in spite of natural selection."[11] They claim that there are further biological forces that, though not "selective," operate in accordance with autonomous nomological generalizations at the same levels of organization as selection, for example, so-called developmental constraints. I have argued elsewhere that no such autonomous generalizations are to be found within biological theory.[12] Even if there were, they would hardly make evolutionary theory probabilistic. Indeed, such constraints might channel evolution narrowly enough to preclude drift. So we shall have to look beyond biology altogether to find the sources of drift.

Let us return to our giraffes. Suppose the naturalists do not launch a careful study of the ecological conditions of the giraffe population, but chalk up the change in gene frequencies to drift, meaning that some *nonevolutionary* forces intervened, ones that operate irregularly, occasionally, in a way not normally correlated with changes in gene frequencies. We are to imagine some Rube Goldberg–like device whose appearance on the scene and whose operations are so singular that biologists can be excused for not accommodating it in their theory. For example, suppose that, through freakish, never-to-be-repeated wind

10. Cf. Richard Lewontin and Stephen J. Gould, "The Spandrels of San Marco and the Panglossian Paradigm," *Proceedings of the Royal Society of London* 205 (1979): 581–598.

11. Ibid., section 5.

12. *Structure of Biological Science*, pp. 242–243. See also chapter 6 below.

conditions, the two tallest trees on which the most attractive vegetation for giraffes grow became so twisted that they accidentally trapped the heads of most of the tallest giraffes and broke their necks. In this case, the change in gene frequencies may be said to be due to drift, but drift will refer to entirely deterministic but utterly nonevolutionary factors, which again through our ignorance we must treat probabilistically. If the naturalists never find these freakish trees and never see in their crooks the ensnared skulls of the poor giraffes, then their attribution of change in gene frequencies to drift will also be epistemic, and again will not reflect any indeterminism in the theory of natural selection.

Of course, we know why the gene frequencies changed, and we have informed the game wardens, who have arrested the poachers. So they have been removed, never to trouble the giraffes again. What are we, who know the facts, to say about the change in gene frequencies? Surely we will not credit the change to drift. We will say that for a short time the environment changed, making long-necks maladaptive and therefore shifting gene frequencies through selection.

If we can generalize from this fictional case, the conclusion seems to be that, from a position of omniscience, there is no need for the notion of drift; evolution simply moves faster among small populations, when their gene frequencies change at all; and the phenomena and the theory of natural selection are thoroughly deterministic, where adaptation, mutation, and migration always operate and are never impeded by any biological obstacle. We finite creatures, however, have need of probabilities when applying the theory and have need of the notion of drift, to serve as an epistemic fig leaf that attributes deviations from expected gene frequencies to the interference of nonevolutionary forces, or protects us from charges or ignorance by enabling us to gather together exceptional and unknown selective forces in the grab bag of drift.

The choice between the two interpretations of drift can be given a terminological gloss. First, we may use 'evolution' to label all change over time in gene and/or genotype frequencies, some of which is adaptive and some of which is not. In that case, adaptational forces and nonadaptational ones (which we denominate 'drift') jointly determine the direction of change. The question for evolutionary biology then becomes, how much of the actual diversity of the biocosm over time is due to selection, to adaptational changes, and how much is due to nonselective changes, changes *uncorrelated* with selection, which are therefore random with respect to adaptation and which we call drift.

If evolution means change, then 'drift' labels those causes of change untreated in the theory of natural selection. To the extent that these causes are conveniently treated by statistical techniques, when combined with the theory of natural selection they provide statistical explanations of change. But then the character of drift does not imply that evolution is in fact stochastic nor that the theory of natural selection is itself statistical.

Second, we may use 'evolution' to name not just any change over time in gene frequencies but only changes in adaptational directions. Then if drift is responsible for any part of evolution, it names those selective causes that operate so irregularly, or about which we know so little, by contrast with mutation, immigration, emigration, mating patterns, and so forth, that we collect them under the name 'drift'. Since they are uncorrelated with other more regular selective forces, the most manageable treatment of them will be statistical, even though the forces 'drift' labels operate deterministically.

By and large, biologists have chosen the first of these terminological alternatives, labeling the course of biological diversification 'evolution' and contrasting selection with drift in its determination. Either way, drift does not make evolution stochastic nor the theory of natural selection statistical, except to the extent that the theory is about us and the way we deal with our ignorance, as well as the changes in evolutionary diversity we hope to explain.

DRIFT AND THE DIRECTION OF EVOLUTION

We can more fully substantiate the view of drift as purely epistemic probability by comparing it with Elliott Sober's arguments to the effect that "*[r]andom genetic drift* . . . is the source of the stochastic element in evolution."[13]

Sober's exposition of the notion of drift is much like Strickberger's. He too describes selection, mutation, and migration as *directional* forces, by contrast with drift. He explains how drift depends on population size, or more exactly, effective population size; reproduces a graph very much like figure 1; points out that, even where the population is large, if it is divided into isolated subpopulations, alleles can drift to

---

13. *The Nature of Selection* (Cambridge: MIT Press, 1984; reprint, Chicago: University of Chicago Press, 1993), p. 110, emphasis in original. Pages in text refer to this work.

fixation and extinction in each subpopulation, while both are still represented in the starting ratio within the whole population.

Then Sober becomes "ambivalent" about the evolutionary role of drift, though not ambivalent in the two ways described above. He writes that, while selection, mutation, and migration determine the direction of evolution, drift seems to determine the *magnitude* of the changes they work. This suggests that drift on the one hand and selection, migration, and mutation on the other are two dimensions of a vector that fixes the time path of evolution.

One of these dimensions is deterministic, according to Sober, and the other is stochastic. Sober claims that the directional forces are deterministic: operating on gene frequencies in the absence of drift, they enable us to predict the expected value of subsequent frequencies with very high probability (p. 111). If this is right, then it is difficult to see how drift could give the magnitude of evolution. Certainly it cannot be viewed as anything like one of the dimensions of a vector. For consider: if drift is zero, then coefficients of selection, migration, and mutation give both the direction of evolution *and its magnitude*. Only when drift is nonzero do these other forces give less. What is more, when drift is very large compared to the other forces, they do not even determine the direction of change in biodiversity. So far from being a component vector, or the dimension of one, it looks more like drift is really a countervailing force. Or perhaps it operates orthogonally to evolutionary forces, in the way that electromagnetism operates orthogonally to mechanical forces.

Of course, Sober recognizes this: "But the vector addition and subtraction that underwrites our understanding of how deterministic forces combine cannot be used here. If drift is a selective force, it is a force of a different color" (p. 117). This admission severely undercuts Sober's distinction between direction and magnitude in apportioning the causal force of selection and drift. We need to take seriously his qualifications, "drift . . . says something about the *magnitude* of change, although it remains silent on its *direction*" (p. 116, emphasis in original). Yes, drift says *something* about magnitude, but then so does selection. And drift is not always entirely *silent* about direction: given small populations, it makes some outcomes more probable than others, even when selection coefficients are high.

This is what I mean by Sober's ambivalence about the role of drift in evolutionary theory. One the one hand, drift is repeatedly described as an evolutionary force (see, for instance, p. 116), and since populations are finite, "there is no doubt that chance plays a role in evolution" (p. 112) (as opposed to merely a role in biological change). On the other hand, it cannot be summed together with other evolutionary forces, like selection, migration, mutation. Drift turns out to look more like a *nonevolutionary force,* or a way of referring to a congeries of such nonevolutionary forces that are responsible for *changes* in gene frequencies but not for their *evolution,* that is, their movement in the direction of greater adaptation to local conditions. On this view, of course, drift is no part of the theory of evolution and cannot be responsible for its statistical character. This is not to deny that it can be combined with a theory of evolution to explain actual diversity. But diversification may not be identical with evolution.

Of course, the other alternative is to view drift, not as a generic label for nonevolutionary forces that intervene to deflect evolution from its course, but as a cover for selective forces of which we are ignorant. This is an approach that makes sense of our appeal to probabilities within the theory itself as epistemic resources: the probabilities reflect our ignorance of what the proportions are among the deterministic forces of evolution, selection, migration, and mutation that are operating. Sober rejects this alternative. And with it, he undermines his claim that drift is "the source of the stochastic element in evolution." At most, he can retain the claim that drift is the source of the stochastic element of our *best theory* of evolution, even though evolution itself is, after all, deterministic. Or so I shall argue.

Is Drift a Useful Fiction?

According to Sober, the reason we distinguish drift from selective forces is to facilitate comparisons among different populations (p. 115), in order to frame interesting evolutionary generalizations, that we would have missed if we hadn't drawn this distinction. "There are *biological* facts that are captured by separating drift from selection that would be *rendered invisible* if this distinction were not drawn" (p. 115, emphasis added). As with similar expressions such as Kitcher's about cytology and Sterelny and Kitcher's about theoretical aims, here too one wants to ask, to whom will these biological facts be "rendered invisible" by declining to draw the drift-selection distinction? Is it just we who would have missed the generalizations, or anyone?

Sober's answer must be that without drift some biological facts will be invisible to anyone, including the omniscient. Sober contrasts his position with what he calls the Laplacian view that the recourse to probability simply reflects our ignorance of the deterministic causes, a view he finds in a passage by Darwin: "I have hitherto sometimes spoken as if the variations . . . had been due to chance. This, of course, is a wholly incorrect expression, but it serves to acknowledge plainly our ignorance of the cause of each particular variation."[14] So, despite his disavowal of drift as an evolutionary force, Sober insists there are *biological facts* that dictate the probabilistic character of evolution. That presumably is why the theory is probabilistic and not simply a reflection of our ignorance. What are these biological facts?

Sober does not cite any biological facts directly. Instead he launches a vigorous argument in favor of autonomy of probabilistic explanations from deterministic ones. That is, he argues that, even when a phenomenon is deterministic, our explanatory purposes may be better served by probabilistic explanations than by deterministic ones: "Even with full knowledge of the details, stochastic modeling may retain its point. Besides excusing our ignorance, the probability concept is an essential one for carving out generalizations" (p. 126).

The argument for the indispensability of probabilistic notions seems to be that evolutionary probabilities are, like most evolutionary notions, supervenient on nonevolutionary properties of organisms, environments, and so forth. This makes possible the identification of classes of events that are homogeneous in their probabilistic causes and effects, even while otherwise quite heterogeneous in their deterministic supervenience bases: that is, there is a common probability that events of one kind will be followed by other events of a common kind, even though we can discover no underlying regularities in the supervenience bases of these events. Thus "the concept of probability allows us to treat this wide range of populations within a single explanatory framework. This is an explanatory advantage of the probability concept that it possesses regardless of whether determinism is true" (p. 126). Therefore, employing the concept of drift, we can frame generalizations that explain the time path of gene frequencies among a wide variety of populations of differing species, at differing population levels in differing environ-

---

14. *On the Origin of Species* (London: John Murray, 1859), p. 131. I suspect that this passage reflects Darwin's conviction, not that nonselective forces operate with selective ones to determine diversity over time, but that chance and selection work "consecutively," in John Beatty's happy phrase—'chance' meaning that the variations on which selection operates are uncorrelated with adaptational opportunities. See Beatty, "Chance and Natural Selection," p. 187.

ments, in common terms. So far we have an argument only for the indispensability of 'drift' *to us.*

Sober asks, what if, underneath it all, evolution is really deterministic, and what if we had all the data and unlimited computational powers? Would we still require probability concepts to carve out these generalizations? If not, he concedes, our use of probabilities would not "mark an objective feature of the world," it would "merely reflect another aspect of our subjective, human perspective" (p. 127), it would be a useful fiction.

Sober's response to this "what if" question is that, even if we had the powers of the Laplacian determinist, the answer would be no, evolutionary probabilities would still be real. But the argument he gives for this claim turn out not to be biological at all, or even to be epistemic or metaphysically fundamental. Instead, Sober argues by analogy from the alleged indispensability of concepts in another theory in another science—one far less well developed than biology—to the "objectivity" of biological probabilities. Sober argues that, even in a Laplacian world, probabilities have all the ontological status that mental states have in a purely physical world. Psychological states, like biological states, are at most supervenient on and not reducible to neurological ones. This should be no surprise: psychological states are biological states. But there is another reason for the supervenience of the mental: psychological states have "content," they are "about" things, they are, as the philosopher says, "intentional"; but there is no way to explain even the supervenience of the intentional on the biological, let alone the physical.

For purposes of carving out generalizations about behavior, Sober argues, science needs to attribute these supervening intentional properties to organisms. Unless we can appeal to the beliefs, desires, hopes, fears, and memories of subjects, we cannot explain or predict their behavior. Therefore, the supervenience of the mental, Sober says, should have no tendency to convince us that there are no such things as psychological states. He concludes, "I suggest that it is equally implausible to think that [supervening] probabilities are unreal in a deterministic universe" (p. 129).

Sober's argument here is breathtaking in the conceptual difficulties and philosophical controversies it vaults over. It is as if he has either forgotten or solved all of the problems of the philosophy of psychology from Gilbert Ryle to Paul Churchland. Many of these difficulties and controversies are the subject of chapter 7. Here I shall only sketch the problems. The reader impressed by Sober's argument should return to this section after reading chapter 7.

Probabilities supervene on deterministic processes, psychological states supervene on neurological processes, true enough. But among philosophers and cognitive scientists there are those who infer from the supervenience of psychological states (1) that they are not real, or (2) that they are not natural kinds that will figure in mature cognitive theory, or (3) that they are merely heuristic devices, useful instruments we employ for predicting the behavior of (other) cognitive agents. These *ors* are not exclusive.

Philosophers and psychologists have adopted these skeptical approaches to psychological properties for two reasons: the first is that we cannot reduce them to physiological states (for complex reasons given in chapter 7); the second is that the scientific indispensability of psychological states is in dispute. The argument for the indispensability of supervenient psychological states is their predictive power with respect to behavior. But given the fact that psychological theory is no better at predicting behavior than common sense, and that common sense hasn't improved in a couple of thousand years, the conclusion that an adequate cognitive theory's appeal to psychological states is indispensable must be controversial.

It is an irony of Sober's argument that, while he is appealing to the theoretical indispensability of psychological states to ground the objectivity of concepts employed in evolutionary biology, philosophers of psychology are appealing to the theoretical indispensability of concepts employed in evolutionary biology to ground the objectivity of scientifically doubtful psychological concepts.[15]

If the vast library of writing in the philosophy of psychology over the last half century teaches us anything, it is that matters are too unsettled in this area to underwrite an argument by analogy from the ontological or methodological status of psychological states. This much is clear: our understanding of psychological states and psychological theory is far less satisfactory than our understanding of evolutionary phenomena and the theory that systematizes them. Thus Sober's appeal to our commitment to mental properties provides no analogical support

15. See, for example, Patricia Kitcher, "In Defence of Intentional Psychology," *Journal of Philosophy* 81 (1984): 89–106. I discuss Kitcher's strategy at length in chapter 7.

for ontological commitment to probabilities in a deterministic world. Evolutionary probabilities and psychological states may both turn out to be instrumental conveniences.

Sober's argument proves what everyone will grant and assumes what is in fact controversial: it is the utility of intentional notions in scientific psychology that is controversial. Their utility in psychology is therefore a poor argument for their reality. The utility of probability in evolutionary contexts is in fact not in doubt. What is in doubt is its interpretation. Are evolutionary probabilities objective properties of events and sets of events *independent of our knowledge of them,* or are they always implicitly relativized to an epistemic background? Are they contingent on available beliefs of sentient creatures attempting to systematize biological processes up to the limits of their own cognitive powers?

Sober provides a final argument for the ontological reality of evolutionary probabilities. Ironically the argument ends up implicitly committing him to the claim that evolutionary probabilities are epistemic.

Two physically identical coins are tossed one after the other. One comes up heads, the other tails. Did they have the same probabilities of coming up heads? In some circumstances we'd say yes, in others no. Which of these assignments is correct? Sober says both are correct, and which characterization we use depends on our purposes. "Neither science nor philosophy provides any general principle for saying whether the two coins 'really' had equal or unequal chances of landing heads" (p. 130). This is apparently because "the probability of an event is relative to a set of propositions" (p. 131). But what are these propositions about? They are in part about evidence, about people's beliefs, I shall argue.

To see this, consider Sober's argument: "[L]ooking . . . at the causal chain leading up to an event, we can see that the probability of the event *evolves.* Relative to what was true at different earlier times, the event may have different probabilities as an objective matter of fact. The same relativity is found when we broaden our perspective and take account of events that do not lie on that causal chain. Other earlier events may confer probabilities on the event; and so may events that are simultaneous with or later than the one in question" (p. 132). This is not right. Probabilities don't change truth values over time. *The probability of an event on a probabilistic causal chain does not evolve,* because for one thing, there is no such unique probability, as Sober has rightly argued. Rather, an event has many different probabilities, each of them with respect to a different set of events on the chain (and off it). The probability of an event *e* with respect to another event distant in time may be lower than the probability of *e* with respect to a nearer one, but it isn't the *same* probability that has grown from small to large over time. A monotonic increase over time in the probability of an event is a sequence of distinct probabilities, each larger than the prior ones. More important, on the assumption of determinism here in force, the only way probabilities can vary from one is if they are *epistemic*. In whatever sense probabilities may evolve, they can do so only relative to evidence, which changes probabilities as we acquire more of it!

Far from being an argument for the existence of probabilities as objective matters of fact in a deterministic world, these considerations show just the reverse. At least they do if by "objective matter of fact" Sober means, as his contrast with Laplacianism indicates, that probabilities are not in the eye of the nonomniscient beholder.

Where does this leave matters? Sober argues that random drift is what makes evolution statistical and that the statistical character of evolution is an objective fact about it, even if the world is deterministic. We have seen, however, that there is some reason to deny that drift is an evolutionary force at all, and this is something that Sober himself grants, or at any rate is ambivalent about. There are at least two ways to accommodate drift to evolution, either as a placeholder for selective forces that we cannot identify, or as a placeholder for nonselective forces. Sober cannot accept the first of these alternatives, for it turns drift into a convenient (indeed Panglossian) fiction, a way of dealing with our ignorance, instead of an independent source of stochastic effects in evolution. He cannot accept the second because it too would deprive drift of its role as such a source within evolution.

So Sober must argue that there are biological facts that the concept of drift captures for evolutionary theory. But we have seen that, in an important respect, if there are such facts they are all (in part) *about us*, for they reflect our ignorance, and our needs in applying evolutionary theory.

IS EVOLUTION STOCHASTIC? IS THE THEORY OF EVOLUTION PROBABILISTIC?

This brings us back to the view of drift as a placeholder. If it is a placeholder for unknown evolutionary forces, then it is at most the source of the statistical character of evolutionary *theory*, as opposed to the source of the stochastic element in evolution itself. And if it is a placeholder for nonevolutionary forces, then it cannot be the source of the probabilistic character of evolutionary theory, for it is not part of this theory at all. The second of these two alternatives makes more sense of the actual claims of biologists about drift. After all, they invariably contrast it with selection. Of course, this behavior of biologists can always be accommodated to the first interpretation, but doing so brings with it the accusation of Panglossianism, and certainly increases the insulation of the theory of natural selection from potential disconfirmation. As a price to pay for explaining the statistical character of the theory, this insulation may seem too high. On the other hand, viewing drift as a placeholder for nonevolutionary forces that sometimes overwhelm the effects of selection not only has no such drawbacks, but is consonant with what we find in other areas of science. Thus no one rejects the inverse-square law of gravitational attraction just because feathers fall more slowly and with far more variable acceleration from occasion to occasion than billiard balls do. The reason is that air resistance, wind, humidity, and other variables, which prevent us from making anything more than a probabilistic prediction of the time it takes a feather to drop a certain distance, are no part of the inverse-square law of gravitational attraction. Mutatis mutandis for the relationship between drift and evolution.

If this view is correct, then it appears to be reasonable to conclude that, like mechanical phenomena, evolutionary phenomena are after all deterministic, or at least as deterministic as underlying quantum indeterminism will allow. But the theory is probabilistic. More exactly, the theory as we actually employ it is. Drift is a phenomenon on which the theory is not *silent*. Simply to treat it as part of what is in effect a ceteris paribus, or imparibus, clause seems as unrepresentative of biological thinking as treating it as a placeholder for unknown selective forces.

In effect, the inclusion of drift in the theory of natural selection is on the one hand an admission of ignorance and on the other an admirable and often successful attempt to improve the theory's powers to predict more accurately and to explain more fully biological phenomena that we observe and that interest us. Drift helps to explain observed diversity that cannot be explained adaptationally. Instead of remaining silent in the face of marked changes that seem to reflect no apparent selective forces, the theory points to epistemic probabilities about the likelihood of nonselective forces' producing reproductive sampling error, which leads us sometimes to expect such outcomes.

There is, of course, a deterministic theory of natural selection in which drift plays no role, but it is either so generic in its claims as to have little predictive content, or so detailed in its enumeration of selective forces—including, for example, the presence of poachers on game preserves—as to be hopelessly unwieldy and beyond our cognitive powers to discover and express.

The relation of the generic version of the theory to the actual version we employ is rather like that of the deterministic or phenomenological version of the second law of thermodynamics to the statistical version, except in two respects: the phenomenological or deterministic second law is false and useful, while the deterministic theory of natural selection is true and (by itself) useless—for creatures like us. The useful theory of natural selection incorporates drift and so is inevitably probabilistic. It too may differ crucially from the statistical version of the second law of thermodynamics. The kind of probabilities it involves are epistemic; they are relative to us, or to other sentient creatures who formulate probabilistic hypotheses on the basis of evidence. This makes our actual theory of natural selection more of a useful instrument than a set of propositions about the world independent of our beliefs about it. It substantiates the sort of instrumentalism about biological theory and the theory of natural selection as applied to our world that the failure of reductionism suggested in chapter 2.

The notion of drift does not fit comfortably into the pure theory and, when made an integral part of it, turns the theory into a set of claims not only about evolution but also about our beliefs. As such, the theory of natural selection biologists employ is very different from other theories, whether deterministic or statistical. But again, this is no reason to challenge the unity of science or the cognitive warrant of biology. It is another reason to endorse the relative instrumentalism of biology.

Biological Instrumentalism and the Levels of Selection

I HAVE ARGUED THAT the failure of reduction and the statistical character of evolutionary theory reflect the fact that biological science is more strongly contingent on the cognitive and calculational powers of *Homo sapiens* than are the physical sciences. The dispute over genic selectionism provides another context in which the same conclusion seems inescapably to be emerging: the kinds that biological theory describes reflect the interests and capacities of biologists, not the way nature's joints are cut.

GENIC SELECTIONISM

Genic selectionism is the thesis that the forces of natural selection select for properties of the genes and not properties of individual organisms, as adaptive. The thesis's initial proponents were G. C. Williams and Richard Dawkins.[1] Williams' motive for genic selection is parsimony, the desire to minimize the kinds of entities and processes to which our scientific theory should be committed. Dawkins' motivation appears to have been a commitment to a sort of reductionism and to sociobiological approaches to animal behavior. Dawkins' and Williams' thesis has not been popular among philosophers of biology. Refuting this thesis of genic selectionism has generally been taken to require showing that selection operates at other levels, and that selection at these levels cannot be explained smoothly by selection at the level of genes. Thus refutations of genic selectionism have either proceeded by or presupposed arguments in favor of the existence of kinds of biological systems at these levels, which are not reducible to the macromolecules that

---

1. See G. C. Williams, *Adaptation and Natural Selection* (Princeton: Princeton University Press, 1966), and Richard Dawkins, *The Selfish Gene* (New York: Oxford University Press, 1976).

constitute living things. Opponents of genic selectionism make common cause, whether their favored level of selection is the genotype, organic individual, population, deme, or species. Much of the debate about genic selectionism has focused on exegetical issues about how its proponents characterize the environment in which traits are selected for: is the relevant environment in which selection takes place the extra- and intracellular milieu of the polynucleotides that make up the gene, or the more usual niche of the organism bearing the genes?[2]

The entire controversy surrounding genic selectionism has an air of unreality about it. Though all parties to the dispute agree that the issue is alleles—post-Mendelian or "classical" genes that come in pairs and code for phenotypes—almost all parties also agree that there are no such classical genes of the sort required to motivate the debate.[3] The debate is about whether the sole locus of selection is the Mendelian allele, but if advances in breeding, physiology, cytology, and molecular biology from 1900 to the present show anything, they show that almost any phenotype in which we are interested is the result of the interaction of the environment and a vast number of diverse and distributed sections of the genetic material, the polynucleotides that make up the body's DNA. The view that we can connect anything like a Mendelian allele or a pair of them with a phenotype of interest has long been stigmatized as "beanbag" genetics.

Mendelian genetic theory still has an instrumental role to play, and a crucial one, in population genetics. As an instrument for enabling us to predict the distribution from generation to generation of certain phenotypic properties, the theory is unrivaled. But as a useful fiction, a heuristic calculating device, the theory will not support a serious debate for or against genic selectionism.

If there is to be a serious debate about genic selectionism, it should not be a debate about whether the Mendelian allele is the only locus of selection; it must be a debate about whether the genetic material, the DNA polynucleotide, however subdivided into functional parts, is the sole level of selection. So understood, genic selectionism has much to recommend it. The only reason to adopt any other view reflects the instrumental character of biology, its character as a discipline whose

2. These exegetical matters have been conveniently sorted out by C. Kenneth Waters, "Tempered Realism about the Forces of Selection," *Philosophy of Science* 58 (1991): 553–573.

3. Even the rare dissenter recognizes something approaching consensus on this view. Cf. Waters, "Why the Anti-reductionist Consensus Won't Survive," pp. 125–139.

kinds are contingent on our powers and interests and are therefore instrumental and not "natural." Keeping in mind that the only way this debate can get off the ground requires an interpretation of 'gene' as the genetic material, let us assess the leading arguments for and against genic selectionism.

GENIC SELECTIONISM AND CAUSAL REALISM

The philosophical argument against genic selectionism was first broached by Elliott Sober and Richard Lewontin.[4] Sober expanded and deepened this argument in *The Nature of Selection*.[5] Subsequently, Sober's arguments were subject to grave criticism initially by C. Kenneth Waters, who was later joined by Kim Sterelny and Philip Kitcher, in defending a modified thesis of genic selection.[6] These three philosophers have defended a thesis that Sterelny and Kitcher label "Dawkinspeak," or *pluralist genic selectionism:* "there are alternative, equally adequate representations of selection processes and . . . , for any selection process, there is a maximal adequate representation which attributes causal efficacy to genic properties" (p. 358). By contrast, Sober's pluralism holds that, in different populations and for different traits, selection may operate at different levels, and at least sometimes it does not operate at the level of the single gene. Waters', Sterelny's, and Kitcher's pluralism allows that, with respect to a single population and a given trait, there are multiple equally true descriptions of how selection operates, and genic selection is always one of them. They contrast this thesis, which they endorse, with another stronger one sometimes embraced by Dawkins, *monist genic selectionism.* According to Sterelny and Kitcher, "[F]or any selection process, there is a uniquely correct representation of that process, a representation which captures the causal structure of the process, and this representation attributes causal efficacy to genic properties" (p. 358). The difference between these two theses turns on

4. "Artifact, Cause, and Genic Selection," *Philosophy of Science* 49 (1982): 157–180.

5. Pages in text refer to this work.

6. C. Kenneth Waters, "Environment, Pragmatics, and Genic Selectionism," *Proceedings and Addresses of the American Philosophical Association* 59 (1985): 359, and "Tempered Realism," pp. 553–573; Sterelny and Kitcher, "The Return of the Gene," pp. 339–362; Kim Sterelny, Philip Kitcher, and C. Kenneth Waters, "The Illusory Riches of Sober's monism," *Journal of Philosophy* 97 (1990): 158–161. Pages in text refer to the Sterelny and Kitcher paper.

the difference between 'maximal adequacy' and 'uniquely correct'. A representation may be maximally adequate compatibly with other representations that are also maximally adequate. A uniquely correct representation cannot share the limelight, whence the contrast between the monism they reject and the pluralism they embrace.

In order to understand Sterelny and Kitcher's claim, let's take it that representations are adequate if they are true, so that a maximally adequate representation is completely correct.[7] Accordingly, a representation can be compatible with other equally adequate representations only if it is incomplete in its description of phenomena, and it will be maximally adequate only if it cannot be made more complete in its description of the phenomenon. A representation of some phenomenon X will be uniquely correct only if it is true and complete in its description of X, so that there is no room for alternative representations of equal adequacy.[8]

Sterelny and Kitcher claim that there are alternative ways of representing selection processes, either as affecting genes or affecting individuals that bear them. Employing their example, we may talk of selection for "genes" for spiderweb building and their selective advantages (understanding such expressions as elliptical for claims about large numbers of varying quantities of the genetic material, which we cannot as yet identify and which may not even be units in the sense of bearing a single molecular structure or even a small number of molecular structures in common). Alternatively we may consider the selective advantages for spiders disposed to web building. Sterelny and Kitcher conclude, as between spiders and their genes, "[t]here is no privileged way to segment the causal chain and isolate the (really) real causal story" (p. 359). To focus exclusively on the webs—properties directly connected with survival and reproduction of the individual organisms that bear the genes—is fallacious, they insist. Doing so begs the question of whether the environment within which selection operates is the individual's environment or the allele's—the individual gene's. "Equally, it is

7. According to several philosophical accounts of causation, such a complete description of the causal structure of a process may be unmanageably complex and lengthy. Depending on the theory of causation adopted, the completely correct description might even have to include mention of the whole state of the universe at each instant of the causal process in question. But this possibility is no reason to doubt the coherence of the notion.

8. A serious but largely philosophical problem lurks here. How do we establish the completeness of a description of any particular phenomenon that includes less than a whole history of the universe? If relations are properties that a complete account must

fallacious to insist that the causal story must be told by focusing on traits of [the genes] which contribute to the survival and reproduction of those individuals. We are left with a general thesis of pluralism: there are alternative maximally adequate representations of the causal structure of the selection process" (p. 358).[9]

Leave aside for the moment the fact that the genetic material can't really be individuated into allelic units of function. Unless there is some reason in principle why a complete representation of the causal structure of the selection process is not possible, pluralism cannot be correct. Sterelny and Kitcher tell us there is no privileged way to segment the causal chain and isolate the (really) real causal story of the selection process. These are two separate claims. Apparently, the reason for the first claim that the causal chain cannot be segmented is that it is equally true to say that phenotypes manifested by individual organisms (or other units larger than the individual allele) are selected for, as it is to say that properties of genes are selected for.[10] So the individual organism's having a phenotype is part of the causal story of its survival and of the survival of genes it contains. And the individual allele's having a certain property is causally necessary for the survival of its lineage and perhaps also for the survival of the individual's lineage. We cannot deny the causal property of being selected for to either of them, and so there is no privileged way to segment the causal chain. But even assuming there is no privileged way to segment the causal chain, this is no reason for the second claim, that we cannot "isolate the (really) real causal story" of what gets selected for.

THE (REALLY) REAL CAUSAL STORY

Why suppose that there is no such *unique* story, no theory that gives *all* the details correctly, and so excludes alternative representations?

---

include, this problem cannot be solved. On the other hand, scientists have little difficulty with this issue.

9. Waters advances the same view: "We can no longer maintain that a true description of a selection process provides a unique correct identification of the operative selective forces and the levels at which each impinges. Instead we must accept the idea that the causes of one and the same selection process can be correctly described by accounts which model selection at different levels" ("Tempered Realism," p. 554). Waters pleads for a pluralism only tenable under a thoroughly nonrealist point of view of the forces of selection.

10. This is Waters' reason for pluralism. On his view, genic and individual selection theories do not conflict once we understand them fully ("Tempered Realism," p. 554).

One might hold that in science there is no such thing, that data always underdetermines theory. Accordingly, we should embrace an instrumental conception of theories in which their adequacy consists, not in truth, but in some other epistemic or practical virtue. Then, of course, it will not be surprising that for different purposes different theories are more suited. In other words, pluralism might recommend itself in biology because, in general, pluralism about scientific theories is in order.

An antecedent commitment to pluralism for scientific theory in general will sustain pluralism about the maximally adequate account of selection and will sustain pluralism about every other phenomenon of scientific interest as well. But in the present connection such a rationale would be impertinent. Sterelny and Kitcher's thesis is not a claim about the nature of scientific theory as such. It's a claim about one particular theory, not every scientific theory. Sterelny and Kitcher need some fact specifically about selectionist phenomena that renders them inaccessible to representations that reconcile truth and completeness. They need an impossibility proof, something like Gödel's proof, which affects the required sort of result in mathematics, or von Neumann's proof that hidden variables are impossible in quantum mechanics as it is currently formulated. But there is no reason to think that anything like this prevails in evolutionary biology. Instead, what Sterelny and Kitcher offer us is something much weaker, something like the Copenhagen interpretation of quantum mechanics—an epistemological thesis instead of an account of the way the world is.

Without an impossibility proof there is no reason to suppose that what they call "the (really) real" story cannot be isolated, at least in principle. Would it suffice for their conclusion that the (really) real causal story is complicated beyond the cognitive powers of mere mortals? What if the best we can do is provide several different stories, all of which are maximally adequate in the sense that we cannot do any better, even though there is a (really) real story beyond our poor powers to express? This view would of course fail to undermine monist genic selectionism as a theory about the world, and it would substantiate pluralist genic selectionism at most as a thesis about us and our powers of discernment and expression. It would underwrite pluralism as a sort of subjectivist second best. Given our feeble powers, the evidence makes

---

But then they should be combinable into a more correct or perhaps even a uniquely correct account, thus undermining Waters' pluralism.

pluralist genic selectionism look right to us, even though to a more powerful intellect monist genic selectionism might suggest itself.

Though my last two chapters give grounds to suppose that this sort of subjectivism is near the truth of the matter, Sterelny and Kitcher do not intend this interpretation of their argument. Their view is that there is no determinate fact of the matter about the level of selection, not just that, whether there is or not, it is epistemically inaccessible to us. They liken pluralist genic selectionism to "instrumentalism" in the philosophy of science: "Pluralism of the kind we espouse has affinities with some traditional views in the philosophy of science. Specifically, our approach is instrumentalist, not of course in denying the existence of entities like genes, but in *opposing the idea that natural selection is a force that acts on some determinate target, such as the genotype or the phenotype*" (p. 359, emphasis added). It is worth asking what Sterelny and Kitcher mean by the phrase I have emphasized: how can selection be a causal force at all if it does not act on some determinate target or other?

One way a causal force might fail to act on a determinate target is when its individual targets are not instances of a single determinate type of target. As a claim about type causation, Sterelny and Kitcher's thesis may simply be that tokens of the type 'selective force' do not act on tokens of any single homogenous type.[11] Sometimes tokens of the type 'selective force' act on 'genotokens' and sometimes on 'phenotokens', sometimes on trait tokens of particular kin groups, demes, populations, species, and so forth. Then there is no single determinate type of target for the forces of selection to aim at. In this case there are no interesting generalizations about the selection of traits. At most, selection targets an unmanageable disjunction of traits of different types of biological systems. This conclusion would be a defense of a default version of pluralism: there is no uniquely accurate representation because all representations are equally uninteresting; there is no single systematic representation of the effects of selection at all—or at least no such generalizations of interest to cognitive agents of our powers.

Such an argument for pluralism bears close connections to Sober's controversial argument in *The Nature of Selection* that disjunctive types are causally inert (pp. 92–96). If a type is causally inert, it is not a

---

11. Recall that tokens are instances of types. Thus the assassination of Lincoln is a token of several types: killing, fatal shooting, assassination, murder, ending of a presidential term of office, and so forth.

natural kind. We understand what it means to say that there are no entities of a certain kind, while accepting the existence of tokens identified by the use of the kind predicate in question. ('Sunset' describes an earth turn; there are no sunsets.) Thus there is no determinate type of target for selection, because the disjunction of types realized by all the particular tokens selected for does not itself constitute some complex disjunctive natural kind. This argument too needs a justification for the claim that disjunctive types cannot figure in scientifically interesting generalizations just because these generalizations would be too complicated for us to express and exploit.

Instead of acting on differing targets on differing occasions, another way a causal force might *fail* to act on any single determinate target is by acting on tokens of several different types at the same time—a bit of genetic material, its organisms, its kin group, and so forth. But this way of not acting on determinate targets provides no reasons to suppose that "the (really) real" causal story cannot be told. It only suggests that, when the uniquely adequate story is told, it will be a complicated one with a lot of interconnections between different levels of organization. Again, this is not a conclusion that sustains pluralism.

Sterelny and Kitcher's appeal to conventionalist theories of space and time as paralleling their own arguments reinforces the suggestion that there is no *biological* basis for their claim that no uniquely right causal story is possible.

> Another way to understand our pluralism is to connect it with conventionalist approaches to space-time theories. Just as conventionalists have insisted that there are alternative accounts of the phenomena which meet all our methodological desiderata, so too we maintain that selection processes can usually be treated, equally adequately, from more than one point of view. The virtue of the genic point of view, on the pluralist account, is not that it alone gets the causal structure right but that it is always available. (p. 359)

This passage suggests that the basis for their pluralism is a general thesis of underdetermination of theory by evidence. But what Sterelny and Kitcher need is at least evidence for the more specific thesis of the underdetermination of *biological* or *evolutionary* theory by evidence.

In the absence of such an argument, pluralism about selection has little to recommend it, over and above the virtues of pluralism everywhere and always.

If we accept that there is in principle a unique maximally adequate representation, a (really) real causal story, the attractions of monist genic selectionism increase. As Sterelny and Kitcher argue, even though with respect to different populations and differing traits there are multiple equally true descriptions of how selection operates, a description in terms of genic selection is always one of these equally true descriptions. But it must be part of the unique maximally adequate representation, "the (really) real" causal story, and since it is the only description that all cases of selection have in common, it must represent a fundamental causal mechanism common to all cases of selection. This is indeed the case.

The only traits that are selected for are inheritable ones, traits that can be passed on to subsequent generations hereditarily. This is true whether the traits are predicated of species, populations, demes, kin groups, individuals, gametes, gene clusters, homologous paired alleles, individual genes, or sequences of nucleotides, for that matter. Traits at each of these levels are transmitted to the next generation only if traits at the next level are transmitted, until of course we get to the level of the genetic material itself, the polynucleotides. Here the buck stops, traits are not selected below this level, because selection does not operate below this level.

In sexual or asexual reproduction the germ-line DNA replicates by template matching; after fertilization, cell division provides further replication by template matching, until at sexual maturity this chain of template-matched DNAs reaches the germ line in the next generation, and eventually reproduces by template matching in the third generation and so on. Throughout this process the replicating, multiplying DNA is producing gene products that interact with the environment—nuclear, cellular, organismal, ecological, and so forth—to produce phenotypes at various levels, which determine the DNA chain's fate in the subsequent generation. The impact of the phenotype on the subsequent generation always depends on the fate of at least one link in the chain of DNA molecules.

This means that, for every property selected at any level above the level of the genetic material, there is another more complicated property of the genetic material that is selected for, whenever the simpler property at the higher level of aggregation is selected for. Adverting to this complex property of the gene tells more of the complete causal story than does adverting to the simpler higher-level property.

For any garden-variety phenotypic-property exemplification we identify as selected for (such as Kitcher and Sterelny's example of web spinning), there is some other complicated relational property of one or more segments of the genetic material that is also selected for. In the case of web spinning, this genic property is the property of coding for the expression of all those enzymes and structural proteins individually necessary and jointly sufficient (within the range of environmental variables to be specified) for web-spinning behavior under conditions where it is appropriate. This complex relational property of the gene is selected for whenever spider web spinning is selected for.

If genic properties are selected for whenever nongenic ones are but not vice versa (as when the gene carries information about its own replicational machinery), then as an account of evolutionary processes, monist genic selectionism has much to recommend itself.

This means that pluralist genic selectionism misleads when it suggests that a representation of selection in terms of genic properties is always merely "available" and at least as good a representation as any other. It's not just "available": from the perspective of *token* causation, genic selection is indispensable to higher-level selection. Without it, no properties at higher levels of organization could ever be subject to selection in successive generations. So the (really) real causal story for selection, *wherever it occurs,* will advert to selection for properties of the genetic material of the germ line, and selection for these properties will provide the ultimate biological explanation for the selection of somatic traits instantiated at higher levels of organization.

TOKEN MONISM AND TYPE INSTRUMENTALISM

The complicated relational property of one or more segments of the genetic material that is selected for whenever some trait of the individual coded for by that gene is selected for may give monist genic selectionism cold comfort. After all, the property selected for will be too complex to be a property of recognizable biological interest. And it will either be so restricted in its instantiation or so disjunctive as to be without systematic significance. So this vindication of monist genic selectionism as a thesis about *tokens* will be of limited methodological interest, if there is a viable alternative to it that can be defended at the level of *types*. In fact, it may be argued against what has been claimed above, that the entire dispute about genic selectionism is about types. No one doubts the thesis of token-genic selectionism. It is genic selection as a thesis about types of genes and individuals that is at stake. It is on theses about types that the prospects for interesting biological generalizations hinge.

I shall endeavor to show that, on the understanding of monist genic selectionism as a thesis about types, it turns out to be more strongly supported. Token-genic selectionism suggests that there is no defensible general theory about autonomous selection above the gene token—the polynucleic acids.

To begin with, we require a criterion for type selection, a test that will tell us whether a trait type is selected for. That is, we need a test we can impose on biologically interesting traits that are repeatedly instantiated by genes, organisms, kin groups, and so forth, to determine whether having the trait is selected for. A biologically interesting trait is roughly a trait we can recognize as the same from instantiation to instantiation, whose incidence is of actual or possible practical, therapeutic, agricultural, industrial, or theoretical interest, as opposed to some gerrymandered artificial trait cooked up to confirm or disconfirm a claim about selection.

In *The Nature of Selection* Sober proposes such a criterion for selection, which he holds to be part of a strong argument against monist genic selection. This principle is important for Sober because it is part of his argument against genic selectionism. The argument proceeds roughly like this:

1. Principle A: a property is selected for if the presence of that property raises fitness in at least one causally relevant background context, and does not lower it in any.
2. If a property is selected for, then the items that exemplify that property selected for together constitute a category, kind, level, or type of unit at which selection operates.
3. Any unit on which selection operates and that is composed of constituent units not selected for (because condition 1 does not obtain) is a natural kind of the theory of natural selection. As such, these units must be noted in any theory of natural selection that does not overlook significant explanatory generalizations.

It is evident that the claim that a property is selected for should be understood as having counterfactual force. To say that a trait is selected for is to say, among other things, that its incidence is a cause of survival and reproductive success in the disjunction of causally relevant background conditions within which it is instantiated, and presumably its persistence over time is not a merely accidental fact unexplained by its adaptive significance.[12] Moreover, if the items whose properties pass the test of principle A are to be deemed causally homogeneous natural kinds, which figure in the generalizations of evolutionary biology, these items' nomological role cannot be based on some merely accidental property they share.

If a trait passes the test of principle A and is a trait properly predicated of items at a level of organization above the gene, genic selectionism must be wrong.

But, claim Waters and Sterelny and Kitcher, principle A can be defended only at the cost of misrepresenting the theory of natural selection. The problem is that there seem to be many properties that Sober, among others, wishes to recognize as selected for but that fail the test of principle A. Ironically enough, Sterelny and Kitcher instance as a counterexample frequency-dependent selection, which has always been a favorite example of Sober himself in other contexts.[13] In frequency-dependent selection a property is selected for until it becomes heavily represented in its population, after which it is selected against. Thus, in the same causally relevant background context, a trait is sometimes selected for and sometimes selected against. The obvious way to defend principle A in a case like this is to split a population into two subgroups, one in which the frequency of the trait is low and its selective effect is uniformly favorable, and one in which the frequency is high and its effects are not uniformly favorable. But "a defense of this kind fails for two connected reasons. First the process of splitting populations may have to continue much further—perhaps to the extent that we ultimately conceive of individual organisms as making up populations in which a particular type of selection occurs" (Sterelny and Kitcher, p. 345). The fact that a process of splitting populations to find subpopulations that are uniform in their causally relevant background conditions *may have to continue* until the subpopulations are singleton sets—individual organisms, or particular genes, for that matter—is no guarantee that it will do so, and is by itself no objection to principle A. As I argued in chapter 2, the biosphere is sufficiently complex that we should not be surprised if no very interesting nomic generalizations about the selection of any particular trait in a large class of cases can

12. This conclusion is the upshot of the distinction that Sober has made famous, between selection for (a causal notion) and selection of (an epiphenomenal notion). See *Nature of Selection.*

13. "Frequency-dependent Causation," *Journal of Philosophy* 79 (1982): 247–253.

be uncovered. Indeed, as I shall argue, if we accept principle A we should be surprised if any of the traits in which biologists interest themselves really are selected for in populations much above the size of the individual organism.

Keep in mind that selection for types proceeds by selection for tokens: on most ontological views, types themselves are abstract. There may be disagreement about what the tokens are and what types they instantiate, but there is agreement on this point: abstract objects don't have causal relations. The direction of selection—for or against—is always determined by the interaction of the token with its particular environment. Over the course of an individual's lifetime, a given trait may be selected for, selected against, or neutral in its selective role, sequentially, as its local environment changes. For a given trait and a given population, the background conditions causally relevant for selection include all the differing environments in which members of the population find themselves. The probability that in every one of these environments a trait of biological interest will not be selected against is vanishingly small. It is for this reason that there are almost no interesting exceptionless generalizations about the traits to be found in evolutionary biology. At most one finds 'tendency' statements, like 'polar species tend to have higher volume to surface area ratios than nonpolar species.' In fact, principle A underwrites and helps explain this paucity of generalizations. Therefore we have an additional reason to endorse it as reflecting the nature of selection.

Sterelny and Kitcher offer a second reason for rejecting principle A.

[A]s many writers have emphasized, evolutionary theory is a statistical theory, not only in its recognition of drift as a factor in evolution but also in its use of fitness coefficients to represent the expected survivorship and reproductive success of organisms. The envisaged splitting of populations to discover some partition in which principle A can be maintained is at odds with the strategy of abstracting from the thousand natural shocks that organisms in natural populations are heir to. In principle we could relate the biography of each organism in the population, explaining in full detail how it developed, reproduced, and survived, just as we could track the motion of each molecule of a sample of a gas. But evolutionary theory, like statistical mechanics, has no use for such a fine grain of description: the aim is to make clear the central tendencies in the history of evolving populations, and to this end, the strategy of averaging, which Sober decries, is entirely appropriate. We conclude there is no basis for any revision that would eliminate those descriptions which run counter to principle A. (p. 345)

In the chapter 4 I examined the argument of this passage, which attributes aims and strategies to a theory, instead of attributing them to theorists who make use of it. As an account of the statistical character of evolutionary theory it is unacceptable. But without its support, Sterelny and Kitcher's objection to principle A has little force, except on an instrumentalist conception of evolutionary theory.

The strongest conclusion about principle A that the statistical character of the theory of natural selection substantiates is this: agents beyond our cognitive capacities who employ it may identify and explain tendencies of evolution that we are not equipped to identify. But there is no reason to conclude that they will miss any facts about evolution that we can explain. The only reason to surrender principle A appears to be that it is inconvenient for us to employ, given our finite powers and the coarseness of the evolutionary detail in which we are interested.

GENERALIZATIONS AND GENIC SELECTIONISM

The only level of biological organization at which principle A has any chance of working is the level of the genetic material. That is, only at this level are selective effects actually uniform enough to pass principle A's test for being selected for.

At every other level of organization, counterexamples to principle A are the stock in trade of contemporary biomedical science. For example, consider a line of laboratory animals raised because they lack a functioning immune response and so make suitable model systems for the study of immunity. Either this is a group within which having a working immune system is selected against (specimens that have one are removed from the breeding population), so that having an immune response cannot be said to be adaptive and therefore to be selected for (which seems false), or we must try to exclude this case from the range of causally relevant background conditions envisioned by principle A.

What argument can be given for excluding these laboratory cases? Surely the argument cannot be merely that these are cases of artificial selection, for that is just a subcategory of natural selection in which the selective force happens to be *Homo sapiens*. It is well known that nature frequently reveals cases in which one species has had selective effects on other species. Consider predator-prey relations, for example. Artificial selection is just another case in which the behavior of one species provides a strong environmental effect on the adaptation and fitness of another. What can happen in the laboratory can happen outside of it.

Can we exclude such counterexamples by invoking the qualification in principle A that implicitly restricts it to the actual environment in which a trait is manifested and not all physically possible environments? A defender of principle A will argue that we should do so, and that we must distinguish the environment from the causally relevant background conditions that work with a trait to effect selection within the given environment. Presumably the molecular biologist's lab is a different environment from the wild and not a legitimate source of counterexamples. Such an approach is ultimately unavailing. If we can distinguish environment from causally relevant background conditions, the resourceful experimentalist can replicate the environment and select against any trait we like. If we cannot distinguish the environment from the causally relevant background conditions in a nonarbitrary manner, then every trait selected for against some causally relevant background conditions will pass principle A. To avoid this trivial conclusion, we have to draw a distinction that returns the force to the counterexamples.

It does not take a science fiction writer to imagine a nomologically permissible background environment that will make any particular maladaptation adaptive, and vice versa. For almost any biologically interesting trait that has been selected for, it takes only a little imagination to dream up perfectly possible Rube Goldberg arrangements that will turn the trait into a maladaptation in at least one causally relevant background condition, even while holding the environment constant. What are clearly excellent adaptations for almost all environments will be turned into serious maladaptation.

This means that, for any interesting biological trait, either that trait actually fails principle A or there is no nomological obstacle to that trait's failing principle A. Accordingly, the claim that some property passes the test of principle A and is selected for may be true, yet lack any nomological force. There will always be some physically possible background condition in which it would maladaptive and so would not be selected. Given Sober's argument for the autonomy of units of selection and its reliance on a counterfactually supportive version of principle A, it follows that there are no units of selection.

Or rather, there are none above the level of the gene. No biologically interesting property satisfies principle A counterfactually construed . . . except, that is, just possibly genic properties involved in gene expression and gene replication! The biochemical properties of the genetic material—the DNA polynucleotide—constitute the sole class of biologically interesting properties that might pass Sober's test, that are selected for in at least some actual causally relevant background conditions, and are selected against in none, including actual and counterfactual ones. The reason is simple. These properties of the DNA polynucleotides are the only ones that are indispensable *in all causally relevant background conditions* to the mechanism of selection on this planet. They are the ubiquitous mechanism of hereditary transmission without which selection cannot obtain.

Though we can design an experiment that will make almost any adaptation into a maladaptation, we cannot do this for significant molecular properties of genes. For example, DNA has the property of copying itself in all causally relevant background conditions where selection operates. We cannot eliminate this property in the laboratory, while preserving the hereditary role of DNA. To eliminate it is to eliminate hereditary transmission. Much the same can be said for the effects of the genetic material in metabolism. Given the ubiquity of the basic processes like oxidative phosphorylation, eliminating the traits of the genetic material in the germ-line cells that allow for the production of the relevant enzyme types in the somatic cells of offspring, simply eliminates the hereditary lineage within which selection is supposed to operate. To the extent that correct protein synthesis in the somatic cells is necessary for survival and reproduction via the germ line, the genetic material's ability to direct protein synthesis will be selected for in all contexts in which there is survival, reproduction, and evolution.

These properties of the genetic material required for gene expression and replication stand a chance of satisfying principle A, while no other properties do. This important difference between biochemical properties of the genetic material and other interesting biological properties is in part a consequence of the supervenience of the latter on the former. In general, where supervenience obtains, the supervening predicates do not figure in discoverable exceptionless nomological generalizations (whence, for example, the anomalousness of the mental: the absence of laws of intentional psychology is a consequence of its supervenience on the neurological). But the properties in the supervenience base can figure in such laws.

Compare apparently smooth reductions: in these cases, the regularities in the reduced theory are shown to relate *indirectly* connected properties as a consequence of more fundamental relations between *more directly* connected properties in the reducing theory. Thus heating up a balloon is an indirect cause of the balloon's bursting because of more direct causal links between the kinetic energy of the constituent molecules and the elasticity of the balloon. The reductive explanation of why the balloon bursts upon heating is thus simple and straightforwardly given by appeal to just a small number of variables that come into play every time the sequence obtains.

In biology, matters are never so simple. Above the macromolecule, every biologically interesting property is realized in virtue of the instantiation of a disjunction of more fundamental properties, which are themselves supervenient on still other properties down to the level of the macromolecule. At each level relations of redundancy and divergence among the properties make for the realization of "higher-level" properties. For example, a given coat color, which has a uniform effect on adaptation in at least some causally relevant background conditions, can be constituted by a disjunction of different pigments, which are themselves the products of a disjunction of enzymes, substrates, and products, most of which have more than one functional role, any one of which is dispensable in producing the relevant coat color, but all of them will turn out to be, at least in part, indirect products of genes indispensable for survival.

Moreover, like any other biologically interesting trait, any particular coat color is not the only "strategy for survival" available to an evolving lineage. In some causally relevant background conditions, it is maladaptive. By contrast, many of the genes whose products are involved in its availability are indispensable in all causally relevant background conditions for all of the available strategies for survival.

Properties of the genetic material responsible for gene expression and gene replication are the only ones that we can have much confidence will not fail the test of principle A.[14] Every nongenetic property

14. Even they may not be invulnerable. Indeed, the existence of the RNA viruses is a constant challenge to the nomological force of the claim that the properties of DNA are everywhere selected for as the carriers of genetic information, or at least not anywhere selected against. That it is not yet an actual challenge reflects the fact that the RNA virus

of biological interest does or would fail principle A. If we are to identify biologically interesting properties above the level of the gene as selected for, we shall need another test, a weaker one than principle A provides. For example, the substitute principle might hold that a trait is selected for if it is adaptive in most causally relevant contexts, or a majority, or some otherwise interesting causally relevant contexts. But installing a weaker test of selection for a trait is tantamount to the admission that claims about selection lack the universality indicative of nomic force. Such a weaker test would be no basis on which to identify natural kinds, units, or levels of selection. Yet biologists do identify traits (genic, individual, kin-group, and even more holistic traits) as selected for. They recognize that such claims about higher-level traits are not fully explained by the selection for types or traits of the genetic material that codes for these individual and other higher-level traits. At most, the individual instances of such traits are so explained.

BIOLOGICAL INSTRUMENTALISM

What biologists aim at is not the explanatory derivation of their rough-and-ready generalizations from laws of a complete theory of the macro-molecules. Rather, they seek the potential completion of token explanations, the completion of explanations of how any particular model system works, ultimately in terms of selection operating on the macro-molecules. Accordingly, claims about the selection for any given trait of the individual organism, kin group, deme, and so forth, should be understood as singular causal judgments, or sets of them. Such explanations are advanced in full recognition of the fact that there is always some causally relevant condition in which the trait will not be adaptive. The explanation of selection for a particular trait in one or more cases does not proceed through the claim that the trait satisfies principle A and is therefore a natural kind, in a causally homogeneous class. The explanation proceeds in an altogether different direction, identifying the particular factors of the local environment that make the trait conducive to survival of the organism and its reproduction. It is only well below these levels that nomic generalizations set in, generalizations about traits of the genetic material that together with descriptions of

---

employs the DNA of a host to survive and reproduce, and that many of the indispensable properties of DNA are also properties of RNA.

the local environment—intra- and extracellular—nominally link initial conditions with the trait instances whose selection we seek to explain.

Why then do we individuate kinds—and seek generalizations—at levels above those of the genetic material? Not because nature carves at these joints but because *we need to do so* and because it's the best we can do. As inhabitants of this biocosm, our interests in creatures extend beyond those of scientific curiosity. Fulfilling these interests requires information different from a theory about the right natural kind categories expressing the true laws of nature. We need to know a good deal about the details of particular set of organisms in a particular niche for reasons of health, nutrition, and so forth. This is information the search for nomological generalizations won't provide. Even the search for such generalizations is conditioned by our own limitations as organisms. Because we have finite calculational and computational powers, we find it useful to pursue the kind of statistical information that Sterelny and Kitcher think is the main subject matter of evolutionary biology—the "central tendencies." This latter fact about us leads us to weaken principle A, to accept that a trait we identify as interesting is selected for just in case, on average, it increases the fitness of lineages we recognize. Because our cognitive limitations determine which tendencies we recognize, combine, and seek explanations for, these traits and lineages will be dictated by practical agricultural, ornamental, nutritional, medical agendas, as well as by our observational capacities. If all we were interested in were nomic generalizations, then monist genic selectionism would be vindicated from an instrumental point of view. The other generalizations about traits of interest to us are either false or reports of local singular causal matters of fact.

To the extent our interests in our fellow creatures—speaking broadly enough now to include all the eukaryotes and all the prokaryotes—are practical and nontheoretical, a focus on genic selection is not convenient, no matter how "adequate" in principle. Representations of selection at other levels will be preferred, and the description of local matters of fact will suffice. This preference has to do, not with cognitive virtues, but with instrumental ones. If we want to preserve a species, it is enough to know in what environments, on average, their traits will be selected against. If we want to improve crop yields, information about averages is enough. If we want to control a pest, it suffices to known which of its traits that directly annoy us are, on average, selected for. Knowing more, knowing the finer details underneath what are to us the "central tendencies," the full story for every token is not only beyond our information-storage powers, but also beyond our computational powers to factor together. Knowing more is also unnecessary, given our interests. These practical interests make the subjective, observer-relative individuation of levels of selection interesting and important. But they do not underwrite the existence of these levels as objective, observer-independent natural kinds above that of the molecular gene and its properties.

Thus we may opt for pluralism about selection, but only as a relational thesis, a claim about the world and our biological theories, which are suitable to the needs and capacities of creatures like us. Our biological science will turn out to be an observer-relative science, an instrument for getting around in our world, and not something the scientific realist will embrace as the best guess as to what the world is really like. This best guess will embody no laws above the level of the genetic material, and perhaps even none at that level. If the nucleic acids prove to be just one among a large number of local solutions to nature's problem of providing the means for hereditary transmission, then genic selection will be local matter of fact, and not a nomological truth.

Theories and Models, Replicators and Interactors

PUT CRUDELY, my argument is that biologists seek and should seek theories that are heuristically valuable because the nomological truths about biological phenomena are too complicated for our weak brains to carry around and make use of. But at what level of organization does complexity require that creatures like us surrender the goal of a complete account of the generalities that describe the way the world works, in favor of useful approximations and fictions that help us survive? Surely the leading principles of the theory of natural selection count as truths and not mere heuristics? Surely the theory that comports them should be given a realist's reading that makes it true everywhere and always, and not just here and for a while?

My answer to these questions is in the affirmative. Surprisingly few philosophers of biology have concurred in the thesis that there is a theory of natural selection in the same sense as there is a theory of Newtonian mechanics. But nothing could more strongly encourage us to treat biological science as an instrumental discipline than the absence of a clearly articulated theory of natural selection. This absence makes my defense of the instrumentalism of biology and my assertion that there is a distinctive theory of natural selection go together uncomfortably. However, my argument for the instrumental character of the rest of biology hinges on the operation of natural selection at the point where diverse physical structures manifest similar effects on their environment. The regularities governing these effects must be nomological. So the thesis that biology is an instrumental science takes hold beneath the level of generality reflected in the theory of natural selection.

The empiricists' hypothetico-deductive approach to the structure of evolutionary theory has been out of favor for a decade or more. Increasingly, philosophers of biology have embraced the so-called semantic approach to theories, according to which a theory is not a set of laws—truths—about the way things are; rather, it is a set of models for which we seek interpretations that enable us to make predictions and to offer "explanations" of particular occurrences. Though I think that the semantic view is unsatisfactory as an account of theory in the physical sciences, the arguments of the last three chapters have led me inexorably to the conclusion that the semantic approach has much to recommend it as an account of theorizing in an instrumental discipline like biology, except, that is, when it comes to the core of the discipline. As an approach to understanding the theory of natural selection, the semantic approach has little to recommend itself and obscures the role that this theory plays in biology.

There is a general phenomenon of natural selection governed by a set of laws that cognitive agents like us could discover and forge into a theory. But that theory has so little content that it is only when we come to apply it to specific phenomena that all the real work gets done. This is where the semantic approach can illuminate the instrumental character of biology.

Before beginning, one apparently terminological issue needs to be made clear. I shall employ the terms 'theory of natural selection', 'theory of evolution', and 'evolutionary theory' as stylistic variants of one another. The theory so variously styled holds that, over the long term, the sole cause of evolution is in hereditary variations in fitness. We must distinguish this theory from the subdiscipline that derives from it, evolutionary biology—a discipline that includes the study of genetic mechanisms for heredity; sources of hereditary variation like mutation, recombination, immigration, emigration, population size, and structure; mating patterns; drift; and so forth, and perhaps even non-Darwinian causes of evolution (such as those Darwin seems wrongly to have countenanced along with natural selection).

Though the assimilation of these terms, 'theory of evolution' and 'theory of natural selection', may appear to be terminological, some philosophers have condemned it. Elliott Sober writes,

> It would probably be helpful in philosophical discussions of evolutionary biology not to use 'the theory of evolution' and 'the theory of natural selection' interchangeably. It would also be useful to be explicit about whether these "theories" are being construed as sets of general propositions that simply describe what would happen in a population if various arrays of forces impinged on it, or as a historical hypothesis that makes claims about what forces have been at work. . . . And . . . it would be profitable

to distinguish "the theory of natural selection" from the single proposition that asserts that heritable variations in fitness implies evolution, if no other evolutionary forces impinge. This proposition circumscribes *one fact* about the consequences of natural selection, but there is more to our understanding of natural selection than this. It is a misleading stipulation to exclude this rich body of theory from "the theory of natural selection."[1]

This passage makes clear why the assimilation of the terms in question is controversial. Much of this chapter, however, is devoted to vindicating the assimilation of 'evolutionary theory' and 'the theory of natural selection' and showing that the assimilation obscures nothing about evolutionary biology while actually *explaining* some of its salient features.

THE "RECEIVED" VIEW

In a widely cited but generally spurned analysis of the structure of Darwinian theory, Mary B. Williams identified five axioms as characteristic of the mechanism of natural selection. In *The Structure of Biological Science* (chapter 4) I attempted to simplify and expound Williams' theory and to show how it organizes research in evolutionary biology.

According to a simplification of Williams' axiomatization, the core of Darwin's theory can be expressed in four of these axioms (the fifth is introduced by Williams for deductive tractability).

1. There is an upper limit to the number of organisms in any generation of a line of biological descent.
2. For each organism there is a positive real number that describes its fitness in a particular environment.
3. If there are hereditary variations in a line of descent that is better adapted to the environment—that is, has a higher fitness number— then the proportionate size of this line of descent will increase over the long run.
4. So long as a line of descent is not on the verge of extinction, there is always a set of relatively better adapted lines of descent whose

---

1. "Fact, Fiction and Fitness," *Journal of Philosophy* 81 (1984): p. 372f. Pages in text refer to this work.

proportion of the population will increase over the long term from generation to generation.

This simplification and the more formal version from which it is derived have several important features. Note that the first axiom stipulates that each ecological niche has a carrying capacity, without specifying its size. Similarly, the second axiom simply asserts that each organism has a level of fitness and that fitness is a relational property with places for the organism and the environment. Again, the theory requires that fitness differences be hereditary, but it is silent on the mechanism of heredity, as well as on which traits are hereditary. The third axiom does not assert that lines of descent with differential fitness actually exist, only that, if they do, certain consequences obtain. The fourth axiom makes the existence claims.

Partly because of the generality of the axiomatization and its silence on so many issues of burning interest to the evolutionary biologist, Williams' account and my exposition have been subject to considerable criticism. In considering the criticism, it is important to keep in mind whether a less abstract version of the theory could meet reasonable criteria for a description of phenomena that is independent of our cognitive and computational powers.

In *The Structure of Biological Science* I defended the adequacy of Williams' axiomatization of evolutionary theory by noting that key Darwinian ideas follow from it. For example, descent with modification and differential perpetuation are easily derived from it, as is the notion that competition between lineages will result in long-term population equilibria, so long as the environment remains constant. Thus, Lotka-Volterra equations for predator-prey interaction, heterozygote superiority, and, for that matter, species selection, all instantiate the sort of equilibrating mechanism that is directly derivable from Williams' axiomatization. I argued that it is a virtue of this axiomatization that it is silent on the mechanism of heredity and that it does not entail any part of Mendelian genetics or its successors. All it requires is that there is hereditary transmission. This is important since Mendelian heredity is a proximal effect of evolution as it has occurred on the earth and not an ultimate cause of evolution anywhere. Similarly, the theory of natural selection should itself be neutral on disputes about the particular course of evolution on the earth and therefore cannot be involved in the disputes between gradualists and their opponents. Darwin of course had views on this matter and was a committed gradualist. But even the most fervent exponent of punctuated-equilibrium theory endorses the mechanism of blind variation and natural selection, which is the cornerstone of evolutionary theory.

Does silence on these issues make the axiomatization vacuous and so undercut its claims to reveal the character of the theory? In evaluating this line of criticism we need to keep in mind what the theory of natural selection should look like if it is to reflect the universality and generality traditionally held to be characteristic of scientific theories. If such generality must be sacrificed for detailed relevance to the day-to-day issues of evolutionary biology, then the prospects that any theory in biology will have the predictive power, explanatory scope, and realistic interpretation common to theories in physics and chemistry must be slimmer than even I suppose.

The most sustained criticism of Williams' axiomatization and my defense of it has been advanced by Elliott Sober, who defends a conception of the evolutionary theory as a "multiplicity of models" (p. 373), instead of an explicitly stabable theory. Sober writes that "Williams' axiomatization has nothing to do with most of the possible causes of evolution that evolutionary theory considers. . . . It does not circumscribe what I would call the *laws* of evolution—the generalizations that describe what would happen to a population should certain forces impinge" (pp. 379, 380). These causes and the generalizations describing them include natural selection, drift, mutation, migration, mating systems, and especially *genetics* and their operation. Sober is, of course, right to say that the axioms identified by Williams as the core of the theory of natural selection do not explicitly mention the heterogeneous members of this particular list of factors, but from that fact it does not follow that it fails to accommodate them and their effects. Thus the words "natural selection" do not appear in axiom 3, but surely its claim that hereditary fitness differences will produce increases in the proportion of the fitter lineages within and between species is the claim that natural selection obtains. Similarly, the theory of evolution requires variation—and thus allows for whatever may cause variation—so it accommodates the causes of variation biologists have been led to seek by Darwinian theory: drift, mutation, migration, differential mating. The causal regularities connecting these factors with variation and natural selection are important components of evolutionary biology, but they are the derived consequences of the general theory, not the theory itself. This point can be made most clearly in connection with the role of genetics in evolutionary biology.

Sober and others have emphasized particularly the fact that Williams' axiomatization omits any mention of Mendelian genetics and its successors, transmission and population genetics.[2] They allege that, since generalizations in evolutionary biology are frequently given an explicitly genetical formulation, the silence of Williams' axiomatization about genetics unsuits it as an account of the theory of natural selection. This criticism is ironical since much of my own motivation for embracing Williams' axiomatization was the desire to avoid mistakes in previous accounts of the theory, which wrongly make Mendelian genetics the core of evolutionary biology.

Why should any particular theory of heredity go unmentioned in a statement of the theory of natural selection? As noted above, the theory requires that at least some traits be hereditary, but it does not require that any particular trait be hereditary, nor does it prescribe or restrict the mechanisms of hereditary transmission. Natural selection works with one or a dozen different mechanisms of heredity. Indeed, if it is a phenomenon not restricted to the earth but at least potentially instantiatable elsewhere in the universe, natural selection cannot require Mendelian inheritance. As it is, earthly organisms employ diploid, haploid, cytoplasmic, and other sorts of inheritance systems. The theory of natural selection can accommodate this diversity of mutually incompatible Mendelian and non-Mendelian mechanisms only by not being committed to any of them.

Moreover, like any terrestrial adaptation, Mendelian and other genetic mechanisms are themselves the results of selection over variation among competing mechanisms for hereditary transmission. The ubiquity of independent assortment and segregation among sexually reproducing organisms shows that they won the competition among hereditary mechanisms for selective success under the peculiar boundary conditions that characterize the earth since sexual reproduction emerged. Had initial conditions on this planet been perhaps only slightly different from what they were, hereditary transmission mechanisms might have been entirely different. Unless the theory of natural selection should be expected to advert to such specific initial conditions as obtained on the earth 4.5 billion years ago, the requirement that it contain any theory of how traits are inherited is seriously mistaken. Or

2. See, for example, Paul Thompson, *The Structure of Biological Theories* (Albany: SUNY Press, 1988); Elisabeth Lloyd, *The Structure and Confirmation of Evolutionary Theory* (Princeton: Princeton University Press, 1993).

else, there is no theory of natural selection worthy of that name, but only a "multiplicity of models."

Just as evolutionary theory needs to be neutral regarding hereditary mechanisms, any particular hereditary mechanism should be compatible with a diversity of non- and anti-Darwinian theories of evolution, and theories of nonevolution, for that matter. It is well known that the Mendelian idea of particulate inheritance was initially hailed by opponents of Darwin as decisive evidence against evolution by natural selection.

At best, requiring the theory of natural selection to advert to genes is like requiring Newtonian mechanics to make mention of the continents. The latter would be a pardonable mistake if the only context in which Newtonian mechanics is applied were the theory of plate tectonics. The demand that genetics figure in the core of evolutionary theory is a mistake obscured by the fact that, until we discover extraterrestrials, the only thing all the subjects of evolution we know about have in common is the nucleic acids.

One reason not to expect any mention of genes or genetic mechanisms in the theory of natural selection should be evident from the discussion in chapter 2. The notion of 'gene' is a convenient heuristic device for systematizing phenomena available to creatures of our cognitive limits. Were we much smarter, we could make do with molecular characterizations; were we less bright, we would have missed even these conceptual instruments useful in our agriculture, health, and other areas of life. Why should a theory of natural processes that is supposed to obtain everywhere and always, make reference to or require expression in terms of concepts that reflect the limitations of *Homo sapiens* and that do not carve nature at the joints?

Every textbook of evolution devotes most of its space to population genetics. So we are inclined to conclude that population genetics must be the intellectual core of the theory of natural selection. Besides, the theory itself, especially as formalized above, is so obvious that one of its greatest defenders said on first reading Darwin's account, "How stupid of me not to have thought of it." Surely we believe there is more to Darwinism than the tautological claim of the survival of the fittest. And surely what more there is must be the intricate algebra of population biology. Not at all.

For reasons noted in the quotation from Sewall Wright in chapter 4, evolutionary biologists employ the notion of 'gene frequencies' as a convenient unit in which to measure the results of various evolutionary forces—that is, factors that make for fitness differences. Because genes provide the unit of measure of the phenomena to which we apply the theory of natural selection, it is inevitable that population genetics will be central to evolutionary biology. But for the same reason, thermometers are crucial to thermodynamics—they reflect proximate effects we can measure, not ultimate causes.

REPLICATORS AND INTERACTORS

Richard Dawkins and David Hull were more conspicuously successful in gaining acceptance for a characterization of the mechanism of evolution in its full generality than either Williams or I was.[3] Let us consider Hull's self-consciously general account of natural selection. Many of the same considerations favoring Williams' account are also reflected in Hull's and Dawkins' more popular presentation. In particular, their presentation reflects the generality of the theory of natural selection while showing why it would have to be complicated even to approach embracing the features that writers like Sober attribute to it.

Hull begins by introducing three technical terms. These notions are functionally characterized, the second and third in terms of the first, and the three suffice to characterize Darwinian evolution in one generalization.

> replicator—an entity that passes on its structure largely intact in successive replications (p. 408)
>
> interactor—an entity that interacts as a cohesive whole with its environment in such a way that this interaction *causes* replication to be differential (p. 408)
>
> selection—a process in which differential extinction and proliferation of interactors *cause* the differential perpetuation of the relevant replicators (p. 409)

According to Hull the theory of natural selection is the simple claim that, where there is adaptational diversity, it is the result of selection. Of course, differences in what evolves as a result of obtaining selection will depend on the initial or boundary conditions within which replicators and interactors emerge. And if in certain environments nothing

---

3. See Dawkins, *Selfish Gene*, and David Hull, *Science as a Process* (Chicago: University of Chicago Press, 1989). Pages in text refer to this work.

evolves, this will be because nothing in these environments acquires the features of replicators and interactors. The claim that all adaptational diversity is the result of selection in the sense defined immediately above seems perfectly general and well enough confirmed to be a good candidate for the status of a law of nature.

Let's consider each of these notions. As Hull defines them, the concepts build on one another. If there are any problems with the first two, these problems ramify in the definition of the last term.

A replicator is defined as an 'entity', that is, a spatiotemporally restricted particular object, a concrete instance or token of some abstract type, and not the type of which the token is an instance. This seems right, because a type or kind is an abstract object incapable of entering into causal relations. A replicator is not the extension of some property or a set of entities of the same structure, for a set does not have a structure to pass on; it has nothing but members. No, a replicator is a particular thing, for example, a DNA polynucleotide that codes for a certain gene product, say insulin. Of course, Hull's definition is self-consciously functional, and lots of things are intended to satisfy it—genes, cells, organisms, demes, populations. But at a minimum the definition surely will have to be satisfied by a particular insulin gene in someone's body. If evolution operates anywhere, it should at least sometimes operate at the level of the gene. If the argument of chapter 5 is correct, it always operates at that level.

Consider the insulin gene in one germ cell of one organism's body at one time or place. Does it in fact satisfy Hull's definition? To begin with, there are the obvious problems of which entity our particular insulin gene is. Is it the nucleotide sequence that gives the primary amino acid sequence of insulin? Is it this sequence plus the sequence of the introns that get snipped out of the messenger RNA before transcription? Is it this sequence plus the sequences of all the other genes whose products are necessary for the biosynthetic pathway from the DNA to the protein—the ribosomal genes, the histone genes, and the genes for catalysts in the production of the proinsulin molecule, and so forth? Is it all these sequences plus the synthesis and repair-mechanism genes that the DNA molecule for insulin needs to persist through time and to replicate? The questions are complicated, yet they must have answers, since we are confident that some combination of these stretches of DNA do amount to an entity that satisfies Hull's definition of a replicator.

The point of these rhetorical questions is that the causal process underlying the apparent uniformity of genes replicating themselves is frighteningly complicated. It is so complicated that it is no easy matter to individuate the entity we want to call *the* replicator, especially if we wish to identify other actual and possible biological systems here or elsewhere in the universe as replicators.

If individuating the gene were much more complicated, it would not be helpful for our purposes in systematizing phenomena to appeal to molecular genes, or for that matter to advert to replicators, for there just wouldn't be an entity we could identify as passing on its structure largely intact. For present purposes let's assume our insulin gene is just one of the polynucleotide sequences—introns and all—that code for the insulin enzyme.

When it replicates, the insulin gene, like all other genes, does so semiconservatively; that is, the gene is a single strand that unwinds from its partner and then serves as a template on which another polynucleotide is formed. This second polynucleotide does not have the same structure as the first: wherever the first has a thymine, the second has an adenine, wherever the first has a cytosine, the second has a guanine, and vice versa. The "offspring" DNA molecule does not share the first DNA molecule's structure at all, let alone "largely."

If by 'structure' Hull means chemical makeup, then DNA molecules don't count as replicators—passing on their structures largely intact in successive replications. But of course DNA molecules are replicators. So we have to interpret the definition of replicator to make the answer come out right. To deal with this inconvenience, we need to give a special understanding to 'successive' replication, so that DNA molecule 1 doesn't count as having replicated until DNA molecule 2 has semiconservatively produced a third DNA molecule with largely the same chemical structure as DNA molecule 1. We need to understand structure in a special way, too. In fact, we need to understand structure as its very philosophical contrary: as function. To begin with, it's well known that there is a lot of redundancy in the genetic code, and that even if DNA molecule 1's replication resulted in a molecule with a quite different order of base pairs in its successors, they would still count as bearing the same structure, provided only that they coded for the same enzyme, that is, provided they had the same function. Plug these two ways of understanding 'successive' and 'structure' into the definition of replicator, and it begins to look like an even more thoroughly functional and more complicated characterization than it appeared in Hull's original definition.

> replicator—an entity that passes along its causal role with respect
> to interactors in replications that count as successive when they
> produce entities with the same function.

Given the original definition, it seemed relatively easy to identify replicators: find one entity, and if before too long there's a second that comes from it and looks pretty much like it, the first is a replicator.

Things won't be this easy in most cases, but in the central ones—like genes—it should be pretty clear what the replicators are. Of course, as noted above, the more we know about molecular genetics, the harder it is to identify the replicators—*the* entities that do the job. Is it the DNA sequence that codes for the enzyme, or is it that sequence (and all others like it, in the case of repeat genes) plus its promoter, its operon, the DNA polymerase genes, repair-enzyme genes, replication genes, and so forth, that count as the replicator? This is roughly the problem of separating the cause from its conditions. We can do it in the case of the molecular gene, but only after we have acquired a lot of causal knowledge.

When we turn to interactors, the problems of connecting the theory to terrestrial phenomena may increase. Like a replicator, an interactor is a particular entity, a token and not a type. It is said to interact with the environment as a cohesive whole—something kinds will have a hard time doing. Kinds don't interact at all. According to Hull, however, the interaction of these tokens causes something that is difficult to describe in terms of tokens: differential *perpetuation*.

What is perpetuated differentially? Surely it is kinds that are perpetuated in differing proportions. After all, no token is perpetuated—every token is mortal. Of course, if we could simply understand the differential perpetuation of types as the differential reproduction of their tokens, Hull's definition of interactor would just be a terminological variant on this.

> interactor—an entity that interacts . . . in such a way as to cause
> differential reproduction of replicators with which it is associated

But this isn't a permissible way to understand 'interactor'. The reason is clear when we examine the definition of selection: a process in which the differential extinction of interactors causes the differential perpetuation of the relevant replicators.

Since replicators are tokens, selection is supposed to cause the differential perpetuation of tokens—particular organisms, not their types (the kind of organism they are), nor their lineages (the line of descent they belong to)—for that matter. But if only types can be perpetuated differentially, the definition of selection is incoherent. How serious a problem is this? Well, if the differential perpetuation of types is the differential reproduction of their respective tokens, then we may reformulate the definition of selection accordingly. Selection will be a process in which the differential extinction and proliferation of inter- actor tokens and their descendants cause the differential reproduction of some replicator tokens and their successors. Which ones? Those associated with the differentially extinguished and proliferated inter- actor tokens.

Suppose that the replicator tokens associated with the interactor tokens do not themselves constitute a structurally homogeneous set; that is, suppose that replicators don't share enough other properties to count as a natural kind. Suppose they don't constitute tokens that interact with their environments in ways regular enough to make for discoverable regularities about these replicators. In short, imagine what molecular biology would contribute to evolutionary theory if the mo- lecular gene on which interaction and replication supervenes were a hundred or a thousand times more complicated and heterogeneous in its physical structure than it has turned out to be. Molecular genetics would not be able to tell us much that is technologically useful or illuminating about how replicators accomplish replication, or why in- teractors are similar to one another in their environmental interactions, and we wouldn't have been able to identify the replicators at all. Then where would we be? Probably at the point of calling the whole thing off, that is, giving up evolutionary theory because the theory of heredity required both to underwrite evolution and to make the theory precise enough to be any more than a just so story is unavailable.

In the light of these details about the nucleic acids and about types and tokens, let's restate the three definitions. We will see how much more complex they have to be than required by Hull's original account, even to approach the level of concrete detail of evolution as it operates here on earth.

replicator—a token that passes on its functional role in the origin and maintenance of interactors to its descendants through cycles of reproduction that count as successive only when the token's functional role with respect to an interactor is repeated by a de- scendant interactor—a token that interacts with its environment as a cohesive whole in such a way that this token causes the rate of replication of the associated replicator token and its descendants to be different from that of other replicator tokens

selection—a process in which differential extinction and proliferation of interactor tokens and their descendants cause the differential perpetuation of the replicator tokens associated with them

Now we have three definitions that are much more complicated than Hull's original ones, but they are definitions realized by molecular genes and by the organisms whose similarity these genes explain and whose interactions have selective consequences. Note that to establish the occurrence of selection requires only the claim that there are interactors or replicators, since each is defined by reference to the other. Everything else, including the existence of replicators and their differential reproduction, follows from the claim that there are interactors and the definition of interactor. Hull tells us that "the axioms of evolutionary theory must include reference to more than the general characterization of replicators, interactors, and lineages, but it need not and cannot include reference to the characteristics of particular replicators, interactors and lineages" (p. 425). It is, however, hard to see how the axioms, if any, of an axiomatized theory of natural selection can say any more about replicators, interactors, and lineages without making mention of such particular characteristics. In doing so, it would surrender its claims to generality and to the nomological status of its axioms.

No statement about a particular genotoken, organism, deme, population, species counts as a law of nature for well-known reasons: laws cannot mention particular places, times, or things. They are supposed to be true everywhere and always. They are supposed to describe the consequences of conditions that could in principle occur anywhere and anytime. As such, they cannot advert to the existence of particular objects at particular times. Apparently the statement that selection occurs qualifies as such a law, as may the claim that adaptational diversity is always and only the result of selection.

Note that selection is defined in terms that are themselves thoroughgoingly functional, without the tincture of a structural commitment. Nothing in the definitions says what a replicator or an interactor needs to be composed of or how it is to be structured in order to fulfill its causal role. Thus, on earth, chains of nucleic acids can be replicators and interactors, and the cells, tissues, organisms, and so forth, that contain them can be interactors, and perhaps replicators too. But on other planets orbiting other stars, or for that matter anywhere else in space where molecules can replicate and the environment can select, other molecules can be replicators and interactors, and evolution can bring forth the wildest dreams of our science fiction literature. The universal applicability of natural selection is what makes the theory of natural selection more than a heuristic instrument for us, more than a set of descriptions implicitly about us as well as the biosphere in which we happen to function. The claim that biology is an instrumental science stops here.

The theory of natural selection has the autonomy from us that theories in chemistry and physics have. It is easy to see why. We can explain, on the basis of physical and chemical principles alone, why natural selection obtains. The process described in chapter 2, in which selection for some effects gives rise to multiple "winners," can now be described in Hull's language: the emergence of evolution requires only that some molecules be both replicators and interactors. If others can figure as interactors alone, then larger combinations than just the original individual replicator molecule will evolve: combinations of both replicator-interactors and simple interactors that enhance the survival of replicator-interactors. Chemical and physical theory, especially thermodynamics and the theory of the chemical bond, will explain how replication is actual. We will need theories of catalysis, and perhaps even enzymology—the theory of the catalytic properties of amino acid chains (and perhaps also nucleic acid chains)—to explain how interactors got started. The key point is that evolution by natural selection is not an emergent mystery, and relies for its occurrence on facts about molecules that obtain throughout the universe. With a certain amount of effort we may even derive the existence of replicators and interactors from the existence of molecules and the appropriate physical theory.

The differences between the Hull-Dawkins account of the theory of natural selection and the Williams axiomatization is largely terminological. The Hull-Dawkins account surrenders terminological simplicity for a reduced number of axioms. It embodies only one axiom: Selection occurs. What it packs into the definition of selection—replicators and interactors—are technicalities Williams avoids at the cost of multiplying the axioms from one to four or five. Undoubtedly there are other ways to express the central laws of the theory of natural selection. What these actual and potential alternatives have in common is the expression of exceptionless universal regularities, laws with all the force of those common to chemistry and physics. At this point biological theory ceases to be heuristic and comes to have the same character— independent of us—as thermodynamics.

FITNESS AND THE CAUSES OF EVOLUTION

Besides failing to mention genetics, according to Sober, Williams' axiomatization is also silent on all those forces that evolutionary biologists seek changes in to explain the occurrences of evolution. Presumably Dawkins' and Hull's accounts suffer from the same defect. Let us consider further how much of a defect this silence really is.

The evolutionary forces are found in what Sober calls "source laws," presumably generalizations about the sources of evolution, such as "natural selection, mutation, migration, breeding structure, random genetic drift, and a few others." Now Sober himself has had something to say about the heterogeneity of items on this list.[4] In fact, other than natural selection, all the other factors also affect variation, on which selection operates. We do not yet know all the sources of variation, and how nature can narrow and broaden the range of variants on which selection can operate. On the one hand, we want a theory to be more than a laundry list of sources, and on the other hand to identify what is common and peculiar to sources in virtue of which they have common effects. In the theory of evolution, what is common and peculiar to the sources of evolution is that they are all channelled through the variable of 'hereditary fitness differences'.

The axiomatization given above is intended to reflect all these forces and others perhaps as yet unknown, in the fitness differences it accords organisms with respect to their environments. Of course, the environment is to be understood as physical and biological, and for any one organism will include its conspecifics, predators, prey, competitors, and so forth. The way the theory accommodates all the heterogeneous factors that make for evolution is by appealing to the fitness differences they all cause. Because the words naming these causes don't figure in the axiomatization, Sober holds that there is no room for the facts they describe in the austere formalization of the theory. Williams and I, on the other hand, argue that the theory should be expected to identify what it is about the diverse causes of evolutionary change that makes

---

4. Consider Sober's distinction between drift and natural selection as components of a vector determining evolution, discussed in chapter 4.

for their common effect. If there are different causes, then the theory does not have the simplicity and elegance of, say, Newtonian mechanics—a theory that finds the common feature of all sources of dynamic change in their quantity of Newtonian force.

Both proponents and opponents of the Williams axiomatization have appealed to comparisons with Newtonian mechanics in order to defend and attack the adequacy of her axiomatization. Sober says that

> Newtonian mechanics provides an array of source laws. The laws of gravitation assert that a gravitational force exists when there are bodies with *mass*. Coulomb's law says that an electrical force exists if there are bodies with *charge*. These source laws show us how we can discover the presence of forces in ways additional to noticing the acceleration of objects. The comparable story is possible in evolutionary theory, except that Rosenberg, following Williams, has not included source laws in the canonical formulation of that theory. (p. 380)

Here I think Sober has given a tendentious reading of the content of Newtonian mechanics in order to suggest that it embodies a variety of source laws about, for example, gravitational forces and electric forces. From this he infers that, to be like Newtonian mechanics, an evolutionary theory must embody source laws. Since the only "source" of evolution in the Williams axiomatization is differences in fitness, it fails to reveal the desired parallel to Newton's theory. But Newtonian mechanics identifies only one source of dynamic change—gravitational force. Electric forces waited another century, along with the strong and weak nuclear forces, to be added to the sources of dynamic change, and they are added in different theories that in fact led to the overthrow of Newtonian mechanics.

Newtonian mechanics is silent on the causes of changes in force and provides no way of measuring force except through its effects on accelerations. This fact is the basis for a long tradition of treating Newton's first law, $F = ma$, as a definition of force, since nothing in the rest of Newton's theory provides a definition, operational or otherwise, of force. By appeal to Hooke's law, which ultimately depends on Newton's laws, about the relation between the extension of a spring balance and mass, we can measure force independently of accelerations and so free $F = ma$ from the charge of vacuity. Hooke's law assures us that one source of force is a compressed spring, and it thus provides a means of measuring force. Plainly, however, the axioms of Newtonian mechanics ought not advert to compressed springs. By the same token, the theory of natural selection should not advert to the multifarious sources of hereditary variation in fitness.

In *The Structure of Biological Science* (chapter 5) I argued that fitness needs to be viewed as a primitive, undefined term in the theory of natural selection, exactly on a par with the role of force or mass as a primitive undefined term in Newtonian mechanics, and for exactly the same reasons. Fitness is measured by its effects because it has so many causes and is supervenient on so many different underlying features of the organism in the environment. (Recall the discussion of supervenience in chapter 2.) Like the notion of gene, a given fitness level in a given environment can be realized by any of an indefinite number of different combinations of traits of the organism and environmental conditions that interact. This means a single quantitative level of fitness cannot be measured by examining *the* organism-environment relation on which it supervenes—there is no one such supervenience base, nor is there a manageable disjunction of such relations. Nor can fitness be measured by its prior causes.[5] Again, this is because the sources of fitness, as Sober would call them, are too heterogeneous to provide a common set of units and a yardstick to calibrate fitness. If there were only a small number of such sources, then there would be source laws of the sort Sober demands an axiomatization include. In fact (as the absence of examples in Sober's discussion should suggest), there are no source laws to be included in evolutionary theory, because anything we can say about the sources of evolutionary change lacks the generality to be called a law. This is exactly what we would expect on the view of biology as an instrumental science.

Since neither the supervenience base nor the causes of fitness can provide the operational measure for fitness, we can only appeal to its effects in population changes to measure it. Now if we elevate the *operational* definition of fitness differences in terms of differential re-

5. It is worth emphasizing here that the fitness differences between individuals and among small classes of organisms in closely similar environments can be assessed individually by appeal to design constraints, for example. But measuring fitness by appeal to its causes in particular organisms in specific environments does not provide the general measure we seek for population biology. For an excellent discussion of such measures, see John Beatty, "Optimal Design Models and the Strategy of Model Building in Evolutionary Biology," *Philosophy of Science* 47 (1980): 532–561. See also C. Kenneth Waters, "Natural Selection without Survival of the Fittest," *Biology and Philosophy* 1 (1983): 207–225; Robert Brandon, *Adaptation and Selection* (Princeton: Princeton University Press, 1990).

productive rates into a stipulative definition, we convert the theory of natural selection into the vacuous claim that organisms with the greatest reproductive rates will have the greatest reproductive rates. Avoiding this consequence is just one of the benefits of treating fitness as a primitive term in the theory of natural selection. More significant is that so viewing fitness enables us to see the methodologically significant parallel between evolutionary theory and Newtonian mechanics.

Both Newtonian mechanics and evolutionary theory have been charged with empirical vacuity and unfalsifiability because exponents of these theories have refused to surrender them in the face of apparently recalcitrant evidence and they have refused to specify in advance the meaning of their theories' explanatory variables. The reason scientists have declined to do either of these things is that both Newtonian mechanics and the theory of natural selection function as central components in the research programs of physics and biology. Such components are unlikely to be surrendered unless substantially improved theories emerge, and such theories will be relatively distant from direct test because of the large number of auxiliary assumptions needed to bring them into contact with experiment and observation.

In the case of Newtonian mechanics, the unreasonable demand of extremist empiricists that all terms be given explicit operational definitions prevents the theory from playing a unifying, explanatory, research-organizing role. By giving its central explanatory terms explicit operational definitions, these treatments of Newtonian theory have restricted its domain of application to only those phenomena where the operations can be carried out. They have exposed the theory to premature falsification when the source of falsity lay in the auxiliary hypotheses that underwrite the operational definitions. Mutatis mutandis the character of its central concepts, as primitive terms undefined in the theory, enables Newtonian theory to play its unifying, organizing role. Max Jammer expresses the situation of concepts like 'force' and 'mass' in Newtonian mechanics with particular clarity.

> No attempts to formalize Newtonian mechanics by a precise and explicit definition of mass have been successful. . . . For these definitions either had to be based on the concept of force as a primitive notion or had to assume a certain dynamical law which explicitly or implicitly involved again the notion of force. . . . Although . . . for medium-sized objects . . . Newtonian mechanics is of the highest degree of verification, its logical structure seems

to defy all attempts at a complete analysis, if it assumed that such
an analysis presupposes explicit definitions of the fundamental
terms involved.[6]

Much of what Jammer says about Newtonian force applies to Darwinian fitness. Because 'fitness' is a primitive in the theory of natural selection, the theory can play its role of organizing the research program of
evolutionary biology, a role recognized in the recurrent charge that the
theory is invincibly "Panglossian" and so unfalsifiable.[7]

FITNESS AS PRIMITIVE AND PROPENSITY

Despite its role in revealing the character of evolutionary theory as a
body of very abstract but undeniably nomological generalizations, the
thesis that 'fitness' is an undefined term of the theory of natural selection has not won complete favor. Many philosophers and biologists
persist in seeking an account of the notion of fitness that combines two
features: it explains why the theory of natural selection is not the empty
tautology that organisms with the highest reproductive ratios have the
highest reproductive ratios, and it accords fitness a definition within
the theory. These philosophers have argued steadily that fitness may be
defined as a probabilistic "propensity" to leave an expected number of
organisms.

The propensity definition of fitness has several things to recommend
it. First of all, there is no doubt that fitness is in fact measured by appeal
to reproductive rates; the particular reproductive rates of individual
organisms are averaged from season to season; to measure fitness of a
group, individual rates are averaged together; and to measure fitness
of a group over time, they are averaged over individuals and then over
seasons. When such relatively raw measures are held to be unrepresentative, these raw data are further manipulated in accordance with varying statistical techniques to provide probabilistically expected rates of
reproduction that are taken to represent accurate measures of fitness.

As with most cases in science, here too measurement proceeds by
applying a scale of units to the effects of the variable being measured.
Thus simple thermometers measure heat by measuring the linear expan-

6. *Concepts of Mass in Classical and Modern Physics* (Cambridge: Harvard University Press, 1961), p. 120.
7. See Lewontin and Gould, "Spandrels of San Marco and the Panglossian Paradigm," pp. 581–598.

sion of liquids in enclosed tubes, because expansion is caused by changes in heat; more complex thermometers measure changes in heat by measuring the diffusion of a gas, the increase in electrical resistance, or the length of a metal rod, none of them causes and all of them effects of changes in heat. Similarly, reproductive rates measure a cause by calibrating its effects. But what is the cause?

If actual reproductive rates are the effect, and if they vary from measurement to measurement the way a random variable does, then there is some temptation to treat their cause as a random variable also. Since actual variables for reproduction often vary around a mean value in well-behaved but evolutionarily nonsignificant ways, it is tempting to treat their causes as probabilistic propensities, and if differences in fitness are the causes of differences in actual average reproductive rates, then fitness must be a probabilistic propensity. One source of the temptation to embrace the propensity definition is that it makes short work of the mistaken claim that, since fitness means reproductive success, the theory of natural selection is vacuous. This callow mistake involves assimilating the meaning of a term to the operation for measuring its value.

Propensity theorists block the error by noting that fitness is a propensity, a disposition, to be distinguished from the occasions on which it is exercised, just as magnetism is a disposition to be distinguished from actually attracting iron filings.[8] Levels of fitness, the probabilistic propensity, cause actual levels of reproduction. That the connection between the probabilistic propensity and actual levels of reproduction is not direct and uniform is just what we would expect. We know there are nonevolutionary interferences in the effects of fitness, interferences reflected by drift, for example. These inferences help sever the conceptual link between actual reproduction and the propensity to reproduce and this blocks the outcome that threatens the theory of natural selection with vacuity.

Here are two examples of proposed propensity definitions of fitness.

The adaptedness of organism O in environment E equals the expected value of its genetic contribution to the next generation,

---

8. Not all propensity theorists do this. Brandon, *Adaption and Selection*, combines the propensity definition with the view that the theory is a definition, so that the interesting issues arise when we ask whether there are any cases of evolution by natural selection. This view is closely akin to the semantic approach to evolutionary theory I treat below.

that is, the sum of the probabilities of leaving a range of possible numbers of sufficiently similar offspring.[9]

The *fitness* of an organism x in environment E equals $n = (df)n$ is the expected number of descendents which x will leave in E, that is, the sum of the probabilities of leaving each of all possible numbers of offspring.[10]

In fact, biologists' actual treatment of the notion of fitness reflects the spirit of these definitions, if not their letter. Thus, Thoday characterizes fitness as *expected* time to extinction—that is, the probability-weighted sum of possible time intervals before the population becomes extinct.[11] Both these definitions must be distinguished from actual representation of a lineage over time, or actual time to extinction for that species. These are only two of many such propensity definitions. As John Beatty and Susan Finsen have pointed out, the proliferation of probabilistic propensities cited to define fitness means that there will be no single probabilistic propensity, but rather a family of them, some of which may be irreconcilable.[12] For example, expected time to extinction conflicts with propensity to leave offspring when the latter propensity results in population levels that exceed carrying capacity and threaten extinction. To solve this problem by distinguishing long- and short-term propensities, so that short-term overpopulation may reduce long-term propensities, simply reiterates the multiplicity of reproductive propensities that the propensity account of fitness will have to reconcile.

This multiplicity of probabilistic propensities is reflected in the notion of 'expectation' that figures in the claim that fitness is an expected value of reproductive levels. The thesis that fitness is a probabilistic propensity requires that the expected value of reproductive rates be the mathematical expectation. If it turns out to be the evolutionarily ex-

---

9. Adapted from Robert Brandon, "Adaptation and Evolutionary Theory," in Elliott Sober, ed., *Conceptual Issues in Evolutionary Biology* (Cambridge: MIT Press, 1984), pp. 58–82.

10. Susan Mills and John Beatty, "The Propensity Definition of Fitness," *Philosophy of Science* 46 (1979): 263–286.

11. J. M. Thoday, *Symposia for Society for Experimental Biology* 7 (1953): 96–113; W. S. Cooper, "Expected Time to Extinction and the Concept of Fundamental Fitness," *Journal of Theoretical Biology* 107 (1984): 603–629.

12. "Rethinking the Propensity Interpretation," pp. 17–30. This paper repays careful study. It includes a number of examples, adapted below, that reflect the difficulty of assimilating fitness to any single or small number of different probabilistic propensities.

pected value, then the alleged virtues of the propensity definition turn out to be merely apparent, as we shall see.

The mathematically expected value of reproductive rates for a type of organism in an environment is the sum of the products of the physically possible levels of reproduction and their respective probabilities. But these probabilities cannot be the de facto relative frequencies over some finite number of past seasons of reproduction. The actual level of reproduction is a reliable indicator of fitness only when the forces that generate evolutionary changes are constant over large enough populations of the generations being counted. When nonselective causes of reproductive changes—like drift—are held constant, changes in actual fitness levels need to be corrected further. Changes in selective forces need to be accounted for in order to calculate expected values of reproduction from actual values. In other words, expected values of reproductive rates are based on evolutionary considerations.

Even when we have made corrected estimates of the probability of each level of biologically possible reproduction for an organism or a group, inferring the fitness-defining propensity is impossible on purely mathematical grounds. Consider the organisms in Table 1 with associated probabilities of reproducing between one and ten offspring. Of course each of these three organisms has the same arithmetic mean for the probabilistic propensity to leave five offspring. If averages give fitness, then all must have the same fitness levels. But it should be obvious that each organism will have a different level of fitness depending on the environment. In an environment with highly efficient predators, it may be fitter to at least sometimes produce ten offspring; in an environment suffering resource depletion, it may be fitter to sometimes have only four offspring. Making corrections to estimates based on these considerations reveals clearly that mere statistical propensities cannot even give uncorrected measures of fitness, let alone define it. Population biologists are familiar with these problems, though they view them all

TABLE 1. Probability of Reproducing Offspring (Percentage)

	Number of Offspring									
	1	2	3	4	5	6	7	8	9	10
Organism 1	5	5	5	20	30	20	5	5	5	0
Organism 2	0	0	0	0	100	0	0	0	0	0
Organism 3	0	0	0	50	0	30	0	0	0	20

as problems of providing a quantitative measure of fitness, not as part of the attempt to define it as statistical quantity.

A possible response to these problems is to adapt my view that fitness is a supervenient concept. I have held that, while fitness goes undefined in the theory of natural selection, it is supervenient on all the occurrent organismal cum environmental factors that causally contribute to those reproductive successes that are cited to measure fitness. Instead, one might argue that fitness is supervenient on a large family of dispositions—the different probabilistic propensities biometricians may uncover. Which probabilistic propensity constitutes fitness over any given period may be hard to determine, but any two organisms that share all the same probabilistic propensities to reproduce over each period—short run, long run, and so forth—will have the same fitness levels, as may two organisms with different probabilistic propensities to reproduce over specified periods. The fact that there is no single or simple probabilistic propensity to which we could definitionally reduce fitness is explained by the vast number of interactive complexity of forces that change the adaptiveness of differing reproductive strategies over time and space.

There are several reasons why an advocate of the instrumental character of biological theory should be sympathetic to the thesis that fitness supervenes on probabilistic propensities. To begin with, it is a tacit admission that no explicit definition of fitness is possible. Supervenience is not a sort of definitional reduction; it is the denial that such reduction is possible. More important, if fitness is supervenient on probabilistic propensities, then it turns out to be just another concept, like 'gene', that reflects our own cognitive and computational limitations, as opposed to some objective feature of the world independent of us. The probabilistic propensities on which fitness supervenes are not the reflection of objective indeterminism in nature. They are not, like the probabilities of quantum mechanics, just the brute expression of indeterminism at work.

We know, from the discussion of drift and the role of population genetics if from nowhere else, that evolutionary biology must have recourse to probabilities even in a world governed by purely deterministic laws, and that quantum indeterminism can explain only the smallest part of biological probability. The only alternative to the pure probabilistic propensities of quantum indeterminism and subjective probability as an interpretation of probabilistic propensities is long-run relative frequences.[13] The gist of our discussion of the relation between actual reproductive rates under various circumstances and expected values suggests that there may be no theoretically useful long-run limit to which relative frequencies tend, and even if there is a limit to long-run relative frequencies, it won't measure explanatorily significant fitness differences as these fitness levels change over the short term in which evolutionary change occurs. This leaves only subjective Bayesian probabilities as the interpretation of probabilistic propensities and therefore as part and parcel of the definition of fitness as probabilistic propensity. If fitness is by definition a subjective probability, then the theory of natural selection is ineliminably subjective itself: it is a theory about the probability estimates of cognitive agents of limited capacities as well as a theory about the causes of biological diversity and adaptation.

This is not an argument I can use. In chapter 2 I argued that it is the blindness of natural selection to causes and underlying mechanisms that makes for the instrumental character of *the rest of* biological theory. Accordingly, I cannot appeal to the instrumental character of the theory of natural selection without undermining the realist's explanation for instrumentalism in biology. Furthermore, the theory of natural selection is not merely an instrument. It describes a general process simple enough to be discovered by cognitive agents of our capacities, has been widely enough confirmed, and is consilient with the rest of science. It is reasonable to infer that evolution by natural selection takes place everywhere and always throughout the universe. The theory satisfies requirements for being a body of exceptionless nomological generalizations about the way things are, and not a heuristic device of some cognitive agents or others. But the theory of natural selection would be merely a heuristic device, if fitness were a probabilistic propensity by definition and not just in the units convenient for us to measure it.

Rather, fitness is best viewed as a primitive term of the theory, undefined within the theory, but supervenient on nonevolutionary nonprobabilistic occurrent properties of organisms in environments and open

---

13. Long-run relative frequencies are the only alternative to epistemic or quantum probabilities unless we adopt a view like Dupré's, according to which our inability to identify strict regularities at levels of aggregation above the quantum mechanical is good grounds to attribute unanalyzable brute probabilistic propensities at these levels and is not otherwise based on more fundamental indeterminacies. See *Disorder of Things*, chapter 9.

to various (mainly probabilistic) measurements as are required on occasion by the theory in which it figures. It is important to bear in mind that to say that fitness is uninterpreted in evolutionary *theory* is not to say that it is given no meaning in evolutionary *biology*. Quite the contrary. Leaving fitness undefined allows for the assimilation to evolutionary theory of all those factors and factors that biologists identify as mattering for adaptation and maladaptation from case to case, but which share in common only their effects on changes of fitness.

Much of the resistance to accepting Williams' axiomatization of the theory of natural selection, or Hull's and Dawkins' for that matter, is diffused when the scope and meaning of the undefined character of fitness becomes clear. The residual feeling that the axiomatization fails to capture the richness of day-to-day evolutionary biology is a misplaced longing for more than any scientific theory can deliver.

MODEL VERSUS THEORY IN THE PHILOSOPHY OF BIOLOGY

The requirement that laws be free from commitment to the existence of particular things at particular places and times has sometimes been felt to be an embarrassment to biology, for all generalizations that figure in it, beyond those of natural selection and organic chemistry, are spatiotemporally restricted—they are about the flora and fauna of this particular clump of matter, the earth. Consequently, biology embodies a large number of rough-and-ready generalizations, statements about tendencies, probabilities, specimens, and so forth, and it suffers from a dearth of exceptionless laws and theories true everywhere and always.

The realization that little qualifies as lawful on this standard has led to several responses: (1) blaming the messenger—denying that every law must be free from spatiotemporal restriction; (2) denying that biology is a fully autonomous theoretical science—it's more like geology, a discipline that embodies no laws of its own but borrows from nomological science; (3) treating the whole issue as terminological—who cares whether the explanatory generalizations of biology meet some logician's criteria, so long as they explain what we need explained in biology.

A good deal of the philosophy of biology over the first two decades of its development reflected concern about this issue. None of the three tactics was fully satisfying. What finally emerged from this debate was a movement to restructure the philosophy of science's account of the nature of laws and theories in all the disciplines. This restructuring was to have as consequences both what was claimed to be a more authentic understanding of actual practice in biology and a better grip on the character of theories and theoretical change in the history of all sciences. The name of this innovation is the semantic approach to scientific theory.

An exclusive allegiance to the semantic view of theory strongly commits its exponents to the claim that biology is an instrumental discipline, though its advocates certainly have not intended this conclusion. In this section I argue for this view.

The semantic approach to theory is contrasted with the hypothetico-deductive, or axiomatic approach, which semantic theorists call a "syntactic" account of theories. According to such view, theories are stated in axioms that express the fundamental laws governing the behavior of their domains and derived theorems that express the laws and regularities explained by derivation from the more fundamental axioms. While the axioms are often couched in theoretical terms, at least some of the theorems are expressed in observational vocabulary, which makes the derived theorems into empirical regularities subject to test. This approach is presumably called syntactic because it makes the relations between fundamental law and derived generalization a matter of logical deduction in the syntax of a given language. Objections to the syntactic view began at this point. It is held that such a characterization binds theories to particular languages, for it is only in a language that deductions can be effected. But the same theory can be expressed in many languages whose derivational characters may differ from one another. Moreover, the syntactic fixation with deductive systems distracts us from concern with the content of theories to the grammar of particular presentations of them, and exaggerates scientists' interest in formal matters of theory presentation, while failing to capture their interest in the content of a theory common to all its alternative axiomatizations. Exponents of the syntactic approach have rejoinders to these objections. Mainly they insist that their approach is on the right track because it provides a means of capturing the explanatory power of theories by its systematization of a large number of derived regularities under a small number of underived laws.[14]

14. It is worth noting that initial proponents of the semantic view eschew the suggestion that there are fundamental underlying laws that it is the objective of science to discover. Instead, the aim of theory is to "save the phenomena." See, for example, van Fraassen, *Scientific Image*.

One real difficulty for hypothetico-deductivism has been accounting for the role of models in scientific theorizing. At least initially, exponents of this view attempted to assimilate the scientist's notion of a model to the logician's, claiming that a model was an interpretation of the abstract calculus of the theory. For various reasons this approach failed, and few now suppose that the syntactic approach can shed much light on the nature and role of models in scientific theorizing. Indeed, one of the consequences of the failure to account for models as interpretations of a partially interpreted axiomatic system was to prune hypothetico-deductivism of its commitments to a postpositivist theory of cognitive significance. According to this theory, correspondence rules give the empirical meaning of the theory's unobservable terms by linking them with a model or interpretation in the logician's sense of a set of objects and extensions. Nowadays, hypothetico-deductivism is just the doctrine that theories can be expressed in axiomatic systems, that the evidence for the axioms is provided indirectly by the confirmation of empirical regularities that the axioms imply as theorems, and that successful axiomatization reflects our best guesses as to what the fundamental laws of working are that govern phenomena in the theory's domain, what the derived laws are, and what the boundary or initial conditions are in describing and explaining observable phenomena.[15]

In contrast to the axiomatic or syntactic approach is the semantic one.[16] According to this approach, a theory specifies, characterizes, or defines an "ideal system"—not a set of fundamental laws—and is embodied by a set of empirical settings in which this ideal system is more or less instantiated. The ideal system defines a state space, a set of dimensions, one for each of the variables whose interrelation the theory describes. Thus a graphic representation of the state space for Newtonian theory about, say, the behavior of eight billiard balls on a table would include one dimension for each of the horizontal and transverse positions of each ball, sixteen in all, plus another eight for the momenta of each of the balls. A graphic representation of a space of twenty-four dimensions like this would require twenty-four axes, all intersecting at an origin. We cannot visualize such a space, and this makes the semantic approach somewhat more difficult to grasp than

15. For a cogent elaboration of this view see David Lewis, *Philosophical Papers*, vol. 2 (Oxford: Oxford University Press, 1986).
16. In outlining this approach I follow the exposition of Elisabeth Lloyd, *Structure and Confirmation of Evolutionary Theory*.

the three-dimensional Cartesian coordinate system we have been familiar with since high school geometry. No matter, spaces of higher dimensionality than Euclid allows have become common in the sciences, physical, biological, and social.

A point in the state space of a theory represents a combination of values of all the variables mentioned in the theory. Thus a coordinate in the state space of a Newtonian theory for a billiard table will contain all possible combinations of twenty-four numbers, in which each set of four gives the position on the table and the momentum of one of the eight balls at a time. The coordinates in the state space for another more complicated theory will have proportionately more numbers in it. The ideal system will contain equations governing the occupation of all the points in its state space. Some of these equations will restrict what point or points can be occupied by the whole set of objects in the theory's domain at any one time—in other words, it tells us what the possible arrangement of objects is at any one time. For instance, no two billiard balls can occupy the same place at the same time. On this view a parameter is a variable we can fix at one and only one value, thus reducing the dimensionality of our state space. In the present example, if all the billiard balls have the same mass, we can restrict the permissible values of the mass variable to a single value, turning a variable into a parameter.

Other equations will govern successive states of the objects in the domain of the theory by specifying which points in the state space are accessible from others. Thus in our Newtonian example, given the position and momenta of the billiard balls at one time, we can fix their positions and momenta at any future time and past time, by applying the theory to calculate permissible momenta transfers. The axiomatic approach would hold that these equations that govern permissible states and the succession among them are the axioms of the theory that reflect the system's fundamental laws of working. Although the semantic approach may accord these equations the name 'laws', it does not take this appellation very seriously.

In one way of presenting the semantic view, a theory is a specification of a state space and a set of equations of coexistence and succession for points in the state space, and *a set of models*. This set of models is not to be confused with the axiomatic notion of an interpretation for an axiomatic system; this is a set of objects themselves, typically sets of mathematical objects, numbers, and sets of numbers, picked out by the equations that describe possible states of coexistence and succession. Application of a theory to explanation and prediction involves seeking sets of empirical data with which these mathematical models are *isomorphic;* that is, they match up one number to one meter reading, so to speak. The essence of the theory is the set of mathematical models that might or might not represent empirical phenomena. As one of the semantic theory's leading exponents, Elisabeth Lloyd, tells us, "Under the semantic view, the general approach to characterizing scientific theories involves defining the intended class of models of the theory. Hence, the theory can be characterized more or less formally, without first defining a set of theorems."[17] Though it is no part of the formal development of the semantic approach, it may be useful to think about its account of theories as follows. In attempting to systematize empirical data, scientists seek mathematical expressions that, for any data inputs, give predictions in the form of numerical outputs that closely approximate the observed values. If possible, they seek to show that the same equations can be employed with data sets from a variety of distinct experiments, observations, and so forth. If they succeed, the class of mathematical models approximately isomorphic with the data sets constitutes the theory, even when no one is in a position to identify the equations that fully characterize the theory. The theory, if expressible, is simply the definition of a complex predicate—one that is usually relational and multigrade—connecting several objects of different logical types. For example, Newtonian mechanics is the definition of "Newtonian particle system" as any system that satisfies $F = ma$, $F = gm_1m_2/d^2$, and $mv_{\text{time } i} = mv_{\text{time } j}$.

In fact, all the action in science is in finding the empirical models and showing that they realize the mathematical ones. The equations relating the mathematical sets simply define the theory. They are not confirmed by empirical data any more than $2 + 2 = 4$ is confirmed by empirical phenomena. Like definitions elsewhere, the interesting question about the equations that constitute a theory is whether they are exemplified by empirical phenomena, not whether they are true or not. As definitions, they are true a priori.

What are the advantages to this approach over the syntactic or axiomatic one? Some philosophers argue that the semantic approach's focus on models more accurately reflects the centrality of models in the actual theoretical practice of scientists and that it also helps us to understand experimental design and the relations between theory and

17. Ibid., p. 15.

data. Because the theory is a set of mathematical models, instead of claims about empirical processes, the further manipulation of these models to make contact with empirical phenomena by auxiliary hypotheses and other methods for controlling inquiry becomes more explicitly prominent. Neither of these arguments is particularly weighty. The importance of models in theoretical science may equally well be accounted for on the axiomatic approach by noting that scientists formulate models as a means of ultimately identifying the fundamental regularities in a domain that can eventually be organized into an axiomatic account; even when scientists know what the underlying laws are, for example in Newtonian mechanical systems, they make use of models in order to circumvent calculational difficulties (e.g., approximation techniques to overcome the three-body problem), and sometimes employ them to make order-of-magnitude predictions when such predictions suffice for various purposes. Few scientists are willing to settle permanently for mere models of a system. They seek ones that improve in predictive accuracy and as a result grow in explanatory power, that is, come successively to provide a better theory of the domain in question. Advocates of the syntactic view have surrendered attempts to analyze scientific models into interpretations of the abstract calculus of a theory's axiomatic presentation. Accordingly, they could embrace the semantic approach's treatment of models, and subsume it under a syntactic approach (as I shall attempt to explain hereafter).

It has long been well known that testing theories requires auxiliary assumptions. The content and role of principles of experimental design are themselves consequences of scientific theory and can in principle be expressed as theorems of an axiomatic presentation of such theory. That diverse theories need to be combined in order to convert empirical data into phenomena—observational regularities that require theoretical systematization—is equally well known and is not an in-principle obstacle to the adequacy of the syntactic view.

A more controversial argument for preferring the semantic view over the syntactic view of theories is its relatively greater attraction for antirealists about theory, who either deny the meaningfulness of theoretical claims about underlying mechanisms, or more moderately deny their knowability and relevance to systematizing data. For instrumentalists about science, the focus on models of empirical phenomena that the semantic approach encourages makes it a preferred alternative to the syntactic approach.

Several well-known features of biological science have been ad-

vanced as considerations in favor of the semantic approach. John Beatty
has argued (along lines similar to mine) that biological theory lacks the
generality of laws characteristic of the physical sciences.[18] On the one
hand, various empirical generalizations one might cite as laws are both
spatiotemporally restricted in their subject matters and riddled with
well-known exceptions that deprive them of nomological status; on the
other hand, Beatty holds, along with other advocates of the propensity
definition, that the one proposition that all might agree bears the requi-
site generality to be a law, the principle of the survival of the fittest, is
in fact a vacuous tautology.[19] The semantic view turns these features
of evolutionary biology from defects into reflections of the adequacy
of the semantic approach.

If the core of a theory is a set of mathematical structures to greater or
lesser extent isomorphic with a set of empirical data from a number
of experiments or observations, then the theory need embody no laws,
or at least we may have a workable theory with explanatory and predictive power long before we uncover the equations to which its "laws"
of succession coexistence give expression. The quotation marks around
the word indicate that these expressions are not laws but definitions of
the complex multigrade relational predicate that constitutes the abstract structure of the theory. This abstract structure is an ideal system
without empirical content. On the semantic view nothing can be found
in the components of a theory that answers to the empiricist notion of
a law—a universally quantified, contingent, counterfactual-supporting,
exceptionless, purely qualitative general statement. And if there is no
such a thing among the components of a theory, and if the nearest
thing to it is a definition, then the theory of natural selection cannot
be faulted for the alleged fact that its central idea is no law, but a
definition. The long-standing problem of explaining why the theory of
natural selection is a respectable scientific theory in spite of the apparent unfalsifiability of its central idea is solved by admitting that this

---

18. See, for example, John Beatty, "Insights and Oversights of Molecular Biology,"
in Michael Ruse, ed., *The Philosophy of Biology* (New York: Macmillan, 1989).

19. For a defense of this approach see Brandon, *Adaptation and Selection,* chapter 4.
Brandon does not explicitly embrace the semantic approach, though he does repudiate
hypothetico-deductivism: "If I am right, the positivist picture of scientific theories is
surely wrong (at least in its application to evolutionary biology). [Footnote:] A large
body of literature exists in philosophy of science in general and in philosophy of biology
in particular on the semantic conception of theories, a new rival to the old positivist
view. My position in this chapter is wholly compatible with the semantic view of theories"
(p. 151).

idea is unfalsifiable, that the core of the theory has no empirical content, is indeed an unfalsifiable definition, but that this is characteristic of all theories—biological or otherwise.

A third reason to favor the semantic approach to biological theory, advanced at length by Paul Thompson, is that, as evolutionary theory is presented in the textbooks, there are at most only infrequent references to laws and theories, but persistent and frequent exposition of models and of ideal mathematical structures that are applied to empirical data.[20] Thompson's argument focuses on the Hardy-Weinberg "law," according to which, in the absence of mutation, migration, differential mating, small population size, and other selective forces, allele ratios remain constant and satisfy the equation $p^2 + 2pq + q^2 = 1$.

Thompson rejects Williams' axiomatization of the theory of natural selection because genetic "laws" of the Hardy-Weinberg sort cannot be derived from it. He rejects Ruse's (*Philosophy of Biology*) axiomatic account of the theory of natural selection because it contains nothing more than Hardy-Weinberg equilibrium as its core and provides no role for selection. On Thompson's view the Hardy-Weinberg "law" is neither an axiom of evolutionary theory nor a derived consequence of the theory, but it is central to the theory. It is just the sort of model we should expect to embody evolutionary theory. This is a view to which Lloyd subscribes. Her book *The Structure and Confirmation of Evolutionary Theory* is devoted to illustrating how the biologist's concern with models, of which the Hardy-Weinberg "law" is the simplest, vindicates the semantic approach as an account of actual theoretical practice in the discipline.

Lloyd rightly notes that population biology is the most formal and developed component of contemporary evolutionary biology. No one denies this, of course. What defenders of the axiomatic approach assert is that we should not mistake the forest of models for the theory that unifies and underwrites them. Nor should we infer from the fact that all the theoretical action is in developing successively more empirically adequate models, that the development of such models is all there is to evolutionary theory. Yet this is what her treatment of the subject suggests. Beginning with Hardy-Weinberg equilibrium, Lloyd describes the succession of models that improve on this model as more and different genetic systems are modeled, in the presence and absence of various sorts of selective forces.

20. *Structure of Evolutionary Theories.*

The Hardy-Weinberg equation, of which [there are] several variations ("single locus models") . . . is the fundamental law of both
coexistence and succession in population genetics theory. As
Lewontin notes, even the dynamic laws of the theory appeal to
only the equilibrium states and steady state distributions, which
are estimated from the Hardy-Weinberg equation or variations
thereof.[21] The Hardy-Weinberg law is a very simple, deterministic
succession law that is used in a very simple state space. As parameters are added to the equation, we get *different* laws technically
speaking (. . . the forms of a model's equations determine its
characteristics). For example, compare the laws used to calculate
the frequency p' of the A allele in the next generation. Including
only the selection coefficient into the basic Hardy-Weinberg law
we get $p' = p/(1 - sq_2)$. Addition of a parameter for mutation
rate yields a completely different law, $p' = p - ps + mq$. We
could consider these laws to be of a single type—variations on
the basic Hardy-Weinberg law—which is usually used in a certain
state space type. The actual state space used in each instance
depends on the genetic characteristics of the system, and not usually on the parameters. For instance, the succession of a system
at Hardy-Weinberg equilibrium and one that is *not* at equilibrium
but is under selection pressure could both be modeled in the same
state space, using different laws.[22]

Starting with deterministic models for pairs of autosomal genes of diploid organisms, Lloyd expounds stochastic variants that give probability distributions for the proportions of genes at subsequent generations,
models that accommodate meiotic drive that shifts the proportions of
the genes from generation to generation in the absence of selection,
models of multilocus genes, structured populations reflecting various
kinship relations, models in which drift plays a role through population
size, founder effects, models of sexual selection, frequency-dependent
selection.

Lloyd concludes, "According to the semantic view, the structure of
a theory can be understood by examining the family of models it presents. In cases of population genetics theory, the set of models—

---

21. Richard Lewontin, *The Genetic Basis of Evolutionary Change* (New York: Columbia University Press, 1974), p. 269.

22. *Structure of Evolutionary Theories*, p. 38. Note that laws should be understood
throughout as simply equations for moving through allowable paths in a state space.

stochastic and deterministic, single locus or multi-locus—can be understood as a related family of models. The question then becomes defining the exact nature of the relationships among them" (p. 38). But presumably there is also the question of stating the exact nature of the relation between these models and the facts about the world that they help us to systematize and presumably to understand. The first question, defining the relation between the models, presumably means specifying the multigrade polyadic predicate, which constitutes the abstract item that all the models interpret. But, as Lloyd and other exponents of the semantic view seem to realize, specifying this predicate is far from easy. Indeed, it appears to be an aim unattainable until all the data is in, and all the best models for each of the data sets have been identified. When we reach this point, the predicate may be so complex that it can perform no heuristic function.

The second question, how is the world arranged so as to make these models useful, is also one to which no illuminating answer may be forthcoming. In the case of mechanical phenomena the answer to the question, what is the world like that makes the models so tractable, is pretty direct: the things in the world really are Newtonian masses, and the relations between them simple enough to be stated even before we have built up a collection of models applying them to data. If the models are as complex and as heterogeneous as they are in evolutionary biology, so that no obvious unification is to be found for them, as exponents of the semantic approach suppose, then this is an admission that, at the level of organization where selection kicks in, nature is too complex, too heterogeneous, and too large a system for our cognitive and computational apparatus to discern the underlying regularities. The world is arranged so as to make these models useful, but how exactly it is arranged to do so is something we cannot describe or even recognize with any nontrivial generality.

If the world reflected no regularities that could explain and unify the models biologists adopt, then there would be no contrast between the way the world really is and the way we are forced to think about it. That is, if there were no exceptionless regularities at even more fundamental levels of organization—those of chemistry and physics— then even these disciplines' current and future shape would reflect our cognitive limitations and not just the way the world really is. This possibility is not suggested by the best evidence so far. The best evidence that there are regularities at the level of physics and chemistry that do not simply reflect our heuristic convenience is their history of relatively

smooth reduction to successively more powerful and broader predictively accurate systems of generalizations. Admittedly, at the current
basement level these generalizations are irreducibly statistical, but the
probabilistic character of quantum mechanics does not, as does that of
biology, turn on epistemic limitations. Rather quantum mechanics reflects a commitment to a world in which the basic regularities are
statistical. But it is also a world in which quite early in the aggregation
of matter the statistical regularities asymptotically approach the deterministic, so that much subluminal mechanics, most medium-range electromagnetism, and almost all chemical synthesis, catalysis, and decom-
position might as well be perfectly deterministic, so far as agents of
even unlimited cognitive power are concerned.

If the universe reflects regularities simple enough for us to discover,
then we can explain the heuristic utility and the formal unity of the
sets of physical and chemical models that the semantic approach identifies as the theories of chemistry and physics. They are useful and share
a common structure because they reflect the regularities that govern
physical and chemical processes independent of us. And the only reason
to be agnostic about this explanation is some antirealist, instrumentalist
scruple about whether our experimental knowledge extends beyond
what we can observe directly or indirectly. To the extent that the semantic approach emphasizes this agnosticism, it has attractions for the
antirealist.

Prescinding from the realist-antirealist debate in the philosophy of
science, it should now be clear why the semantic approach holds special
appeal in biology. To begin with, the prominence, not to say exclusivity, it gives to population genetics models reflects the widespread conviction that genetics is what the theory of evolution should be all about,
since gene frequencies play so central a role in the explananda of the
theory. If population genetics is what evolutionary theory is all about,
and population genetics is a sequence of models, then evolutionary
biology is all about that sequence of models. And if the only construal
of claims about variation and selection makes it into the definition of
a predicate for which each of these models provides an interpretation,
all the better. For it has always been suspected that the theory was
vacuous. If there is no synthetic, contingent theory systematizing the
phenomena of evolutionary biology, then either biology is a profoundly
different science from the physical sciences, or we need entirely to rethink our appreciation of these latter sciences as well, or both!

Instead of these radical alternatives, better simply to recognize the instrumental character of biology—no very profound difference from a philosophical point of view, and no difference from a methodological one. Then we can still help ourselves to the models employed to apply evolutionary theory to biological phenomena at levels of description convenient to us. Better to hold that a body of exceptionless nomological generalizations work together with the regularities of physical science to bring about and to explain the behavior of the biocosm. Then, beneath the level of generality of the theory of evolution, antirealism about biological theory is dictated by the way the world is. On a realistic construal of what our theories tell us about the world independent of us, *we* cannot hope to frame anything other than the sort of models the semantic view holds up as the objectives of theoretical science. If we are realists about physics and chemistry, we may even accept the physical models the semantic approach offers us, but retain our axiomatic accounts as best guesses about they physical and chemical facts that make the models useful. We cannot identify an axiomatic account reflecting simple explanatory regularities about the biological, which will explain the usefulness of our models, because there is none. Rather, there is none we can grasp, because the simplest regularities about the biological are beyond our cognitive powers. And the biological regularities that are within these powers do not reflect the way the world is independent of us. For that reason they are captured, not by any axiomatic account of the way the world is, but by a series of models we find useful in dealing with the world as we find it.

Instrumental Biology and Intentional Psychology

P SYCHOLOGY IS A biological science, for the systems with which it deals—human and infrahuman—are biological systems. Like biology, psychology is a discipline that has faced many problems traditionally identified as philosophical. Indeed, many of these problems parallel the problems of the philosophy of biology, though the most serious of them antedate the problems of biology by literally centuries. After all, it was Descartes who established in the seventeenth century the obstacles to a biological study of the mind. Ironically, many of the advances in the philosophy of biology owe their origins to advances in the philosophy of psychology, which continues to this day to grapple with the obstacles Descartes placed before the biological study of the mind. But while the philosophy of biology draws inspiration and illumination from the philosophy of psychology, many philosophers of psychology have begun drawing methodological and conceptual morals for psychology from biology itself and its post-Darwinian advances. These appeals to biology usually figure in attempts to defend the scientific integrity of psychology in the face of conceptual challenges to its scientific character and prospects, by showing its similarity to biology.

If, as I have argued, we need to reconsider the character of biology and begin treating it as an instrumental discipline, then we shall have to do so as well for psychology, for two reasons. First, if psychology is a compartment of biology, then it is perforce an instrumental science, or rather, part of one. Second, if methods and concepts of psychology are to be defended by establishing their parallel to those of biology, then the heuristic status of the latter must influence the former. What is more, psychology will come to be seen as even more of an instrumental discipline than biology is: its generalizations will be of even more limited scope than those of biology, and its prospects for uncovering general theories about behavior will be at least as remote as biology's prospects for uncovering general theories about the physiology on

which behavior supervenes. If psychological properties irreducibly supervene on biological ones, and biological ones irreducibly supervene
on physical ones, then neither will figure as natural kinds in laws of
nature, and psychological concepts and regularities must perforce be
even more conditioned by human cognitive limitations than those of
biology. Or so I will argue in this chapter.

Common sense embodies an implicit psychological theory, which
explains human behavior largely as the result of beliefs and desires,
which jointly determine our actions. Beliefs it accounts for by their
connection to our sense organs and memory; desires are bodily in origin
but are also the result of the interaction of beliefs. At a minimum,
psychological theories formalize, quantify, and extend the folk psychology of common sense in many ways, and sometimes, of course, they
contradict it. The most important thing contemporary psychological
theory has in common with folk psychology is a commitment to what
philosophers call 'intentionality': to the claim that psychological states
of both humans and infrahumans have "content," they are "about"
actual and possible (and, for that matter, impossible) states of affairs.
For example, when I think of Paris, my thought is "about" Paris; if
you and I both believe that Paris is the capital of France, what makes
our beliefs the same is that they are about the same thing—they both
"contain" the same thought, expressed in my language as "Paris is the
capital of France" and perhaps in yours as "Paris ist der Hauptstadt
Frankreiches." For scientific psychology the terms 'about' and 'content'
are in scare quotes to indicate that they are metaphors that have to be
cashed in by a psychological theory that ultimately explains how our
brain states have content. Here controversy emerges. Since the time of
Descartes, philosophers and psychologists have despaired of providing
an account of intentionality—the idea that thought has content—in
terms of neurological states of the brain.

Among these despairing philosophers and psychologists, especially
in the twentieth century, some have concluded that no such reduction
of the psychological to the physiological is possible, that the commitment to it is an obstacle to scientific advance in psychology, and that
the discipline must surrender the explanatory appeal to the intentional
just as biology had to surrender the explanatory appeal to vital forces
and God's intentions. These writers, formerly behaviorists, now eliminativists, point to what they claim is the explanatory sterility of folk
psychology and its predictive weakness, and diagnose both these disor-
ders as stemming from an unscientific commitment to "intentionality."

The weakness of folk psychology consists in (1) the wide range of psychological phenomena for which it has no account and (2) the fact that, no matter how good one thinks its predictions (and some exponents of intentional psychology think it a very successful predictive instrument), it has made no significant improvement in predictive powers as far back as recorded history will take us.

For many psychologists and philosophers of psychology, forgoing intentionality is unthinkable. Most of these recognize the obligation to provide a scientific explanation of how one chunk of matter (a collection of brain cells) can be "about" another chunk of matter (say, Paris). Psychologists and philosophers who take this view are often called naturalists. They are optimistic about the prospects for a "naturalization" of the intentional. This is where the appeal to biology and to its philosophy enters. The naturalists' most attractive answer about the physical basis for intentionality exploits the same strategy that biology employs to explain how one chunk of matter (say, the giraffe's neck) became adapted to another chunk of matter (its environment): variation and selection.

The first step is to recognize that the content of a brain state is reflected in its effects (behavior), just as the "content" of a given fitness level is represented in its effects (reproduction). Moreover, action (i.e., intentional behavior) is evidently a form of teleological behavior; it is action undertaken to attain some aim, end, or purpose. The successes of a Darwinian approach to teleology have rightly stimulated psychologists of all types to follow suit: to provide a purely causal account of complex psychological processes in terms of selection, variation, and transmission of more basic ones. Indeed, some philosophers have even argued that this is the only course psychological theory can follow. Thus, bringing together three distinct camps in psychological theorizing, Daniel Dennett has written, "Cognitive theoreticians may proceed . . . fruitfully . . . but if they expect AI [artificial intelligence] to pay their debts some day, they must acknowledge that the processes involved will bear the analogy to natural selection exemplified by the [Skinnerian] law of effect."[1] I suspect that, even if we substitute for "artificial intelligence" a far broader term, like "biology," this is a view many philosophers and psychologists would still embrace. The reason is that psychological processes are teleological and there seems to be no alternative to natural selection as the causal source of the teleological.

---

1. *Brainstorms* (Montgomery, VT: Bradford Books, 1978), p. 86.

Without a nonteleological underlay, scientists are rightly dubious of the probity of any teleological theory. So there is an almost irrepressible motive to pursue a natural selectionist account of the psychological. The most efficient way for psychologists to do this is to consciously exploit the apparently successful strategies of evolutionary biology.

There is another motivation for defending the intellectual integrity of intentional psychology by showing its parallel to evolutionary biology. Like folk psychology, evolutionary theory is often taxed with predictive weakness. If it can be shown that, despite the truth of this charge, the theory of natural selection remains respectable science, then intentional psychology can excuse its own alleged predictive weakness on a sort of tu quoque argument from evolutionary biology.

INTENTIONALITY AND THE ANALOGY TO EVOLUTIONARY THEORY

Philosophical exponents of intentional psychology have adopted one or both of two strategies: (1) showing that the weaknesses of intentional psychology are shared with a perfectly respectable scientific discipline, evolutionary biology; (2) showing that intentional psychology shares the explanatory strength of biological theories, because the ascription of intentional content bears important analogies to the ascription of evolutionary adaptation or fitness.[2] These two strategies will make common cause if the absence of nomological generalizations in both evolutionary biology and intentional psychology can be traced to the same basis—the supervenience of fitness and of intentionality.

An especially clear and cogent version of the first strategy, which tars evolutionary biology with the brush of psychological theory's weaknesses and concludes that the latter must be promising since the former has been established as scientifically significant, is provided in Patricia Kitcher's "In Defence of Intentional Psychology."[3] Kitcher de-

---

2. Daniel Dennett, "Intentional Systems in Cognitive Ethology," *Behavioral and Brain Sciences* 6 (1983): 343–390 (reprinted in *The Intentional Stance* [Cambridge: MIT Press, 1987]) illustrates both these strategies aptly. Defending intentional psychology against the charge of theoretical weakness, he concedes that "[a]daptationalism [in evolutionary biology] and mentalism [intentional psychology] are not theories in one sense. They are stances that serve to organize data, explain interrelations, and generate questions to ask nature" (p. 353). Exploiting the strengths of evolutionary biology for intentional psychology, he writes, "When biologists ask the evolutionists' why-questions they are, like mentalists, seeking the rationale that explains why a feature was selected" (p. 354).

3. Pages in text refer to this work.

fends intentional psychology against the charge that it is without predictive content, roughly by admitting the charge and then going on to
show that evolutionary biology is without predictive content either.
She concludes, "The moral to draw [from biology] for intentional psychology is: when the phenomena under study exhibit this sort of complexity, do not expect to be able to apply general laws in explanation
or prediction, and do not expect to be able to offer any dramatic predictions for real cases" (p. 105). Accordingly, we are not to expect predictive power in a theory describing phenomena of the complexity of
the psychological. Nevertheless, such reduced expectations are scientifically responsible. Though they may be respectable, such expectations
will be problematic unless we have a cogent explanation for them, not
just a rationalization from the unexplained limitations of biology.

Kitcher's conclusion rests on an account of what she calls the "transition laws" of population genetics. These are, broadly, the Mendelian
principles that govern hereditary transmission. One problem with her
argument is that, as we have seen for other evolutionary generalizations, these are subject to controversies. There is no agreement about
whether these principles are evolutionary generalizations, or even about
exactly what the generalizations of this discipline are. If we follow
Philip Kitcher, we must hold that neither Mendel's laws nor any successor of them should be viewed as nomological. If Mendel's laws and
their successors are at best heuristic devices, convenient for cognitive
agents like us, it will hardly vindicate intentional psychology's laws to
show they are no different from the generalizations of genetics.

There are, as Patricia Kitcher notes, practical problems surrounding
the employment of these principles to produce predictions about terrestrial phenomena that have anything like the impressiveness of, say, the
existence of a new element or the occurrence of a solar eclipse at a given
time. Kitcher invokes Lewontin's (1974) explanation of the difficulties.
These surround the specification of initial conditions needed for any
attempt to generate evolutionary predictions. As she says, "[T]he state
descriptions needed to carry out the program of predicting the genetic
composition of later populations on the basis of initial conditions and
laws of transformation are just not available" (p. 104). It is crucial to
note that this explanation of the predictive weakness of population
genetics grants the geneticist knowledge of the "transformation laws"
(or their first approximations) to be applied to state descriptions. It
also identifies the problem of specifying initial conditions as due to the
complexity that any measurement of the "fitness of a genotype" must

face. Extra- and intragenomic interactions on which evolutionary fitness depends cannot be ignored, held constant, or aggregated in a way needed for significant prediction. In effect, evolutionary fitness cannot be measured, given present resources, and so no firm evolutionary predictions can be made. Mutatis mutandis for intentional psychology. Its explanatory variables cannot be measured with the accuracy required for predictions of actual behavior, and therefore it too must be plagued by predictive weakness. We have already seen how attractive these considerations have made the semantic approach to evolutionary biology. Are transformation laws already known in intentional psychology, and does the difficulty of measuring intentional states lie in the complexity of the components of these generalizations? Unless the answers to these questions are on balance affirmative, the problems that bedevil evolutionary biology will provide little support for the prospects of intentional psychology, even on the dubious assumption that we have uncovered laws of transmission.

The obvious candidates for transmission laws in intentional psychology are the generalizations we can abstract from ordinary explanations of human action that have been given the rubric "folk psychology." Such generalizations connecting intentional states with one another and with behavior are widely held to be at least the first approximations to the laws of a scientific psychology. One such a generalization might take the following form:

L—If an agent has desire d, and believes that available action a is the most efficient means of attaining d, then the agent will do a.

The causal variables that L and other generalizations of folk psychology advert to, beliefs, desires, recognition, consciousness, perception, and actions, are supposed to be among the natural kinds of an intentional psychology.[4] The predictive weakness of folk psychology is explained by pointing to problems in measuring these folk-psychological variables identical to the ones bedeviling the causal variables of evolutionary theory.

Here the first strategy for defending intentional psychology makes common cause with the second. The second strategy is showing that intentional psychology shares the strengths of evolutionary theory, not the weaknesses of evolutionary biology. The argument in brief proceeds

---

4. This view is propounded forcefully by Terry Horgan and James Woodward, "Folk Psychology Will Always Be with Us," *Philosophical Review* 94 (1985): 197–226.

as follows: evolutionary theory "naturalized" design (adaptation, fitness) by explaining it as the product of hereditary variation and natural selection. Similarly, we may "naturalize" intentionality, bring it within the ambit of the biological, by showing that these states have their intentional content as a result of the operation of selective forces over *neural* variation, both phylogenetic and developmental. The most sustained development of this approach to the intentional is advanced by Ruth Millikan in *Language, Thought, and Other Biological Categories.*[5] But the locus classicus of this naturalization of the intentional is Daniel Dennett's *Content and Consciousness.*[6] Intentional states are thus no more and no less acceptable to a naturalistic conception of behavior than fitness and adaptation are. But now return to the first strategy for defending intentional psychology: both fitness and intentionality are the products of diverse forces of selection operating on highly complex products of variation; thus neither can be identified or measured in a way that permits precise prediction before the fact, or explanation after it.

Evolutionary naturalism thus bids fair to let intentional psychology have its cake and eat it too. When the characteristics of intentional content are assimilated to those of biological fitness, the irreducibility of psychology is ensured along with its scientific respectability—it's just like biology. As a result, psychologists and philosophers feel confident in their answer to psychological eliminativists like B. F. Skinner and philosophical ones like W. V. O. Quine. Eliminativists pose a dilemma predicated on the claim, originally made by the late-nineteenth-century German philosopher and psychologist Brentano, that psychological intentionality cannot even in principle be explained physically. In Quine's widely quoted words, "One may accept the Brentano thesis

5. Other exponents of this view are Daniel Lloyd, *Simple Minds* (Cambridge: Bradford Books, MIT Press, 1989); Fred Dretske, *Knowledge and the Flow of Information* (Cambridge: Bradford Books, MIT Press, 1981), and *Explaining Behavior* (Cambridge: Bradford Books, MIT Press, 1988).

6. Dennett's exploitation of this view continues in *Brainstorms, Intentional Stance,* and *Consciousness Explained* (Boston: Little, Brown, 1991). Those already acquainted with Dennett's views about intentional psychology know that his early works were decidedly sympathetic to an instrumentalist treatment of intentional psychology. See, for instance, "The Intentional Stance" in *Brainstorms.* It is important to note, however, that the evolutionary analysis of intentionality that he pioneered in *Content and Consciousness* is independent of his initial instrumentalism, and that he has retained a commitment to the selectionist approach to intentionality while apparently surrendering instrumentalism about intentional states. See "Real Patterns," *Journal of Philosophy* 88 (1991): 27–51.

either as showing the indispensability of intentional idioms and the importance of an autonomous science of intention, or as showing the baselessness of intentional idioms, and the emptiness of a science of intention."[7] Quine chose the second of these alternatives because "the true and ultimate structure of reality . . . knows no propositional attitudes, but only the physical constitution and behavior of organisms" (p. 221). Quine went on to adopt an instrumental view of the intentional concepts we employ to describe and explain our behavior in everyday life. He calls such statements as "essentially dramatic idiom," suggesting that much of what we say employing these terms is false but employed to give vent to our nonintentional states and to affect the behavior of others.

Quine's views and those of eliminativists have never been popular. But their arguments from the alleged predictive weakness (or at least unimprovability) of intentional psychology and our inability to understand how physical states could by themselves bear content, have given empiricist philosophers and psychologists serious problems to solve. The assimilation of the intentional to the evolutionary makes them free not to have to choose between intentionality and physicalism. The selection on variation that results in both adapted organisms and intentional states relies on nothing more or less than "the physical constitution and behavior of organisms" broadly considered. Accordingly, an intentional science may, like biology, also be irreducible and indispensable, even though it too concerns only physical constitution and behavior.

The aptness of analogies between intentional psychology and evolutionary biology for defending the integrity of the former thus turns in large part on the nature and significance of that biology's irreducibility to the rest of science. If evolutionary biology is independent from nonevolutionary science, and if intentional psychology is irreducible to nonpsychological science in the same way, then exponents of intentional psychology will be able to have their cake and eat it too. Few will reject evolutionary biology just on the strength of its irreducibility to the rest of science—especially when it has been explained in a way satisfactory to exponents of the unity of science.

By parity of reasoning from biology, many can accept the sort of parallel irreducibility that an evolutionary naturalization of the inten-

7. *Word and Object* (Cambridge: MIT Press, 1960), p. 221.

tional provides. Kitcher exemplifies this strategy as much as she does the first.

> The objection [to intentional psychology] would be that since [it] is irreducible to a more basic science, it cannot be properly integrated into the rest of science. . . . An obvious reply to make in defence is that this science is not unique in being irreducible to a more basic science. For the last ten years, philosophers of biology have been arguing that transmission genetics is not reducible to molecular biology. In this case, the failure of reducibility has raised doubts about the adequacy of philosophical views about the relations among theories. A pressing item on the agenda for philosophy is to elucidate important relations among theories besides reduction.
>
> This work is just beginning, but it still seems clear that there are interesting relations to be explicated, in which case, the quick inference from irreducibility to the impossibility of theoretical integration would be blocked. (p. 102)

Philosophers of biology have uncovered a general feature of the relations between transmission genetics and molecular biology, and between fitness ascriptions (to genes, genotypes, or organisms) and theories about the underlying structure of the genetic material. This mereological dependence in biology consists in supervenience: while there are no manageable type-type identities between the kind terms of these theories, biologists are committed to token-token identities between the items that satisfy these theories.

Token-token identities enable the biologist to extend evolutionary and/or genetic explanations of particular phenomena in a piecemeal way that reveals the particular phenomena's underlying mechanism without committing the biologist to the operation of the same underlying mechanism in another token of the same evolutionary or genetic type. For example, the persistence of sickle-cell anemia as a case of balanced polymorphism can now be explained all the way down to the substitution of one nucleotide in the DNA sequence for hemoglobin.

The extension of the original evolutionary explanation that discoveries in molecular biology provided does not suggest, however, that other cases of balanced polymorphism have any very similar underlying mechanism. Other tokens of the type 'balanced polymorphism' have underlying mechanisms of different molecular types, some of which we already know in as much detail as the mechanism behind sickle-cell anemia. Thus the "reduction" of evolutionary regularities about balanced polymorphism, for example, will always be piecemeal unless the number of different molecular mechanisms that realize balanced polymorphism is far smaller than we have reason to believe. Accordingly, evolutionary theory, and transmission genetics for that matter, can be expected to remain irreducible to more fundamental theory, at least for cognitive agents of our capacities.

There is prima facie reason to think that the supervenience relations that obtain in evolutionary biology are mirrored by intentional psychology. Fitness supervenes on organism and environment. Intentional states too supervene on neural structures and the intentional agents' environments. Because different neural states and different environments can produce intentional tokens of the same type—for example, the belief that Paris is the capital of France—the irreducibility of psychology to nonintentional science is on a par with that of the irreducibility of evolutionary to nonevolutionary biology. Of course, philosophers who have sought to exploit this parallel have not noticed that the autonomy of evolutionary biology is a reflection of its instrumental status. But mutatis mutandis, if biology is an instrumental science, then intentional psychology is one as well, for two reasons. First, psychology is a compartment of biology, and second, intentional terms are not type identical to any nonintentional neuroscientific terms we can bear in mind.

As noted in chapter 6, fitness levels and differences between organisms are measured in terms of their effects in differential reproduction. Like many theoretical variables, for instance temperature, the fitness of an organism is not measured or identified by reference to its causes or the structures that realize it from case to case, but by reference to its effects. A given level of fitness is individuated by rates of subsequent reproduction, suitably corrected. These numbers and units are the "content" of a fitness level. Two different organisms in two quite different environments with two quite different fates may have exactly the same fitness levels, if and only if they have the same (suitably corrected) rates of subsequent reproduction. The fact that this content is supervenient on nonevolutionary properties of organisms and that it is given by the very effects that fitness levels are called upon to explain is the source of both the irreducibility of evolutionary biology and its predictive weakness. If, contrary to fact, the number of properties on which fitness supervenes were small enough for us to manage, then type-type identities between fitness levels and nonevolutionary properties of organisms would be practically possible. Such type-type identities might enable us to determine numerical fitness levels without prior knowledge of their effects in differential reproduction. As a result, evolutionary theory would be predictively stronger. Alas, nature is more complicated than would suit our convenience.

Intentional states are in the same boat. Indeed, philosophers of psychology built the boat; it was they who invented the contemporary notion of 'supervenience', only to have philosophers of biology appropriate it for their purposes.[8] These states, of which the paradigms are states of belief and desire, are also identified by their effects on behavior. They cannot be identified by appeal to the neural states on which they supervene, or the environmental stimuli that help determine their content. Intentional states are individuated by their content; what makes the belief that Paris is the capital of France different from the belief that London is the capital of Great Britain is the difference in the propositions they contain. And the content of an intentional state is identified by an inference from its effects on behavior, often verbal behavior. Identification of particular intentional states proceeds only by attributing a large number of other collateral intentional states to agents. In this respect the attribution of a particular intentional state is rather like attributing a level of fitness to a particular property of an organism, like its neck length, or its speed, or its molecular structure.

Attribution of fitness to particular structures or behaviors (as opposed to the whole organism) involves estimating, or at least holding constant, the fitness levels of the other adaptively significant components of the individual organism.[9] Since in both the evolutionary and the intentional case such estimates or ceteris paribus assumptions are

8. See Donald Davidson, "Psychology as Philosophy," in *Essays on Action and Events* (Oxford: Oxford University Press, 1981). 'Supervenience' is a notion invented for moral philosophy by R. M. Hare. The first exploitation of Davidson's development of the notion of supervenience in the philosophy of biology is in Alexander Rosenberg, "The Supervenience of Biological Concepts," *Philosophy of Science* 45 (1978): 368–386.

9. Indeed, the parallel is more extensive. Among evolutionary biologists are those who condemn the identification of anatomical structures as having specific adaptational significance, on the ground that such structures do not face selection individually but only in the company of the rest of the organism. This makes indeterminate the ascriptions of adaptational "content" to a part of the organism, since a different ascription together with other adjustments in our adaptational identifications can result in the same level of fitness for the whole organism. In the philosophy of psychology, the dual of this thesis is reflected in the indeterminacy of interpretation. Cf. Lewontin and Gould, "Spandrels of San Marco and the Panglossian Paradigm," and Davidson, *Essays on Action and Events*.

rarely correct, attributions of particular levels of fitness, or of intentional content, based on subsequent behavior are likely to be highly inexact and subject to continued refinement and revision. This inevitably gives the explanations of folk psychology and evolutionary biology the appearance of "Panglossianism" and empirical weakness: nothing by way of evidence can count against these theories; evidence can only discredit our estimates of initial conditions of their application.

The analogical argument from evolutionary biology for giving intentional psychology a chance thus appears to have considerable strength. But what if intentional psychology turns out to be related to the rest of biology the way that biology is related to the rest of natural science? That is, if psychology turns out to be a heuristic device relative to a theory that is itself a heuristic device, then psychology's predictive powers are going to be even weaker than biology's. The claims of its kind terms to divide nature at the joints and reveal the nature of thought will be substantially weaker than biology's claims to reveal the nature of biological phenomena.[10]

NATURALISM IN ACTION

Let us consider the details of two kindred attempts to naturalize the intentional, those of Daniel Dennett and Jonathan Bennett.[11] By now

10. For those acquainted with the philosophy of psychology, it is worth explaining why the discussion hereafter is silent on the theories of Jerry Fodor. First, Fodor rejects the sort of evolutionary-inspired naturalism under discussion, and he does so for reasons similar to my own, though apparently arrived at independently. Second, Fodor is so satisfied with the predictive power of folk psychology that he considers its generalizations to be laws, albeit *inexact,* and sees no need or expectation that they be linked to those of other sciences. Thus, unlike the writers treated here, he lacks a motive to explain away the weakness of intentional psychology. Third, Fodor feels no need to show the unity of intentional psychology with nonintentional science. He therefore also spurns naturalism in this broader sense. Fodor is a physicalist—he does believe that the wet stuff in our skulls has intentional states and represents the way the world is. Unlike naturalists of the sort dealt with here, Fodor believes that there is an innate language of thought, a basic intentional symbol system hardwired into neurology, out of which more complex thoughts are built. Thus the intentional content of our thoughts is not contingent on the environmental appropriateness of our neural states. Rather, the environmental appropriateness of these states is a result of their intentional content. See Fodor, "A Theory of Content: Part 1," in *A Theory of Content and Other Essays* (Cambridge: MIT Press, 1990), chapter 2. I suspect that at some point Fodor must come face to face with the problem of how the basic hardwired intentional states emerged in our ancestors. Given the appeal to hardwiring, there will be no alternative to an adaptational account, and no escape from the problem of explaining intensionality that naturalism faces.

11. Daniel Dennett, *Content and Consciousness* (London: Routledge and Kegan Paul, 1969); Jonathan Bennett, *Linguistic Behaviour* (Indianapolis: Hackett, 1989). Pages in text refer to this work by Dennett.

venerable, these theories have been extremely influential, and the new wrinkles added to them by philosophers like Millikan, Dretske, and Lloyd have little bearing on the features of naturalism that they illustrate. These approaches suggest that the identity of a psychological state is given by its causal origins and/or its subsequent effects, and that the individuation of such states proceeds by identifying their effects, in the same way adaptations in nature are uncovered. As Dennett puts it, the adaptive character of human behavior is a product of the evolution *of* the species' brain, and the content of a brain state is a product of evolution *in* the brains of individual members of the species.

> [S]ince environmental significance is extrinsic to any physical features of neural events, and since the useful brain must discriminate its events along lines of environmental significance, the brain's discriminations cannot be a function of any extensional, physical descriptions of stimulation and past locomotion alone. Rather some capacity must be found in the brain to generate and preserve fortuitously appropriate structures. . . . [A] close analogue of natural selection of species would be a system that could provide this capacity and could itself be provided for by natural selection of species. (p. 72)

On Dennett's theory,

> The content, if any, of a neural state, event, or structure depends on two factors: its source in stimulation, and whatever *appropriate* further efferent effects it has. . . . The point of the first factor in content ascription, dependence on stimulus conditions, is this: unless an event is somehow related to external conditions and their effect on sense organs, there will be no ground for giving it any particular *reference* to objects in the world. . . .
>
> The point about the link with efferent activity and eventually with behavior [the second factor] is this: what [a neural] event or state 'means to' an organism also depends on what it does with the event or state. (p. 76)

Variation and selection among nascent neural connections in the developing brain of an organism provide for the buildup of afferent-efferent channels connected at the afferent end to the environment and at the efferent end to the movements of the organism's body that reflect "environmental appropriateness" of the neural events in the channels. During growth and development certain genetically hardwired afferent-efferent channels are selected for by their environmental adaptation; they compete with and extinguish other channels, and finally come to be bundled together in ways that are mirrored by their effects in smooth coordination of bodily movements. These movements have precise enough environmental appropriateness to justify highly specific ascriptions of content to the neural states "inside" the organisms that cause them. The details of Dennett's scenario are compelling, though difficult to summarize. They result in a convincing demystification of psychological content: if Fido eats in the presence of steak, this reflects an afferent-efferent channel from food's presence to mastication. We can see exactly how and why some of the neural events in the channel have the content whose "referent" is the piece of meat and others whose meaning is "eat the meat!" or something like it.

At this point complications set in. As Dennett points out, when it comes to identifying the precise content of Fido's neural states, there will be difficulties, for the dog presumably doesn't have the concept of steak, or butchered animal part, or even food: "What the dog recognizes this object as is something for which there is no English word, which should not surprise us—why should the differentiations of a dog's brain match the differentiations of dictionary English?" (p. 85). But do they match up with the differentiations of any dictionary?

In presence of the steak that Fido goes on to wolf down, the contents of some of Fido's neural states include some statement or other about environmental stimuli, as well as a statement about his subsequent behavior. But they include no statement in particular about either stimuli or behavior, at least no statement we can frame in English. To do so involves attributing doxastic states and preferences to Fido about steak, or beef, or meat, or butchered animal part, or food, or some coreferring term or coextensive predicates of English, Urdu, Hausa, or Chinese, and Fido knows no language. Nor is his behavior so precise, nuanced, and stereotyped that we could attribute a specific belief or desire to him that is describable in any one of these languages.

But if we cannot ascribe a particular set of contents to his doxastic states and preferences, we cannot make any very specific predictions about his behavior, or explain it after the fact except in the most general terms. If we cannot do either of these things, naturalism and the psychological theories it inspires are at a scientific dead end. Even in the case of organisms whose behavior gives a much more specific signal of intentional state, like humans and their talk, behavior is not enough of a guide to intentional states to enable us to improve on the explanations and predictions of folk psychology.

The problem of ascribing a very specific belief or desire to Fido, or whatever intentional states are suitable to dogs, also bears on animals with language. If naturalism is to be a recipe for an intentional psychology that is a scientific improvement on or a successor to folk psychology, it will have to provide a method of identifying the determinants of behavior more precisely than folk psychology does. Insofar as it remains methodologically on a par with evolutionary theory, it may not need to be able to specify initial conditions inside the head in terms or units that are independent of subsequent behavior: it may continue to measure causes by their effects. But it will have to do a better job of this than folk psychology currently does.[12]

This problem makes Jonathan Bennett's version of naturalism particularly attractive. It holds out the prospect of attributing quite specific statements to the intentional states of organisms, both linguistic and nonlinguistic. Indeed, Bennett is particularly interested in solving the problem of such attribution to nonverbal animals, because his aim is to produce a theory of linguistic behavior by "upgrading" and combining nonlinguistically naturalized intentional states into ones finally complicated enough to endow their subject with language. The details of Bennett's strategy are perfectly compatible with Dennett's; in fact they compliment that theory perfectly. While Dennett's is an ontogenic, developmental version of naturalism, Bennett's is a synchronic version of the same theory, providing an account of intentionality across the phylogenetic spectrum, instead of down the evolutionary lineage.

Bennett is particularly interested in developing a theory of beliefs, in the reasonable expectation that the naturalization of beliefs will provide a model for and a building block in the account of other intentional states. We will follow him in this expectation, and hereafter focus on the naturalization of beliefs. He begins with a notion he invents and implicitly defines, 'registration': organisms register propositions about their environments. But registration is not 'perception' or any other particular intentional state. What they register is a function of what

12. Of course, an instrumentalist version of naturalism, like Dennett's view in *Content and Consciousness*, will not have to provide for such improvement, since it in effect denies that intentional psychology is an improvable scientific theory. Rather, it is a heuristic instrument already exploited up to its maximum level of precision in ordinary life.

their goals are, where goals are not intentionally construed, but are read off behavior in the way we read the goals of nonintentional teleological systems.[13] Like Dennett, Bennett anchors registration in environmental input and behavioral output (p. 54) but doesn't tell us anything about it beyond its causal transactions: "[Registration is] a theoretical term, standing for whatever-it-is about organisms which validates predictions of its behavior from facts about its environment; comparable to introducing 'gene' to stand for whatever-it-is about organisms which validates predictions about offspring from facts about their parents" (p. 52). (Though his interests are not biological, Bennett's parallel to 'gene' is apposite in the present connection, for intentional states are supposed to be like the subjects of genetic transmission laws. For that matter, Bennett might have likened registration to fitness, and later in his book explicitly does so.) Organism a registers that P, if "a is in an environment which is *relevantly similar* to some environment where P is *conspicuously* the case." The italicized expressions are shorthand jargon: each must be defined nonintensionally.

At first blush Bennett's theory seems to be quite unlike Dennett's. The latter is a theory about neural content, about the way in which afferent-efferent channels and small groups of them come to have content. Bennett's theory seems to be about states of the whole organism. In fact, both theories come down to the same thing. As we shall see, what an organism registers, on Bennett's definition of the term, is only what relatively small portions of its neural equipment register.

Bennett upgrades registration into belief as follows: 'a believes that P' is true if a registers that P, and a is highly educable with regard to many kinds of propositions that do not include any important kind to which P belongs. Educability is defined as the conjunction of two purely behavioral properties of organisms: organism a is educable with respect to proposition P if

1. a registers that P in environments $K_1, \ldots, K_n$, that is, a evinces behavior appropriate to P's being the case in these environments. Suppose, however, that P is not the case in K_n. Then eventually a ceases to evince P-appropriate behavior in K_n. The less time this takes, the more educable a is.
2. a registers that P in environments $K_1, \ldots, K_n$, but not in K_{n+1},

---

13. Bennett offers a powerful analysis of teleology that shows how goal directedness is the result of wholly mechanistic processes, in the case of the adaptativeness of both hereditary and nonhereditary features. See Bennett, *Linguistic Behaviour*, chapter 2.

although P is the case in K_{n+1}. Then eventually a begins to evince behavior in K_{n+1} appropriate to P's being the case in it. The less time this takes, the more educable a is.

So belief is educable registration. The mechanism of educability may well be variation, selection, and packaging that Dennett envisions as basis of complex intentional states in neurologically sophisticated organisms.

Of course, Bennett's proposal for identifying the contents of educable registrations in the absence of a language that we can attribute to the subject of such registrations, faces exactly the same problems Dennett recognized.

> We cannot [by using ordinary language] express a's beliefs in terms which exactly capture his epistemic state. In fact, our belief-attributions will usually credit him with too much. . . . But we allow ourselves such inaccuracies in attributing beliefs to one another, so why should we struggle heroically to avoid them in attributing non-linguistic belief? Let us do what we conveniently can to avoid inflating these attributions, always remembering that we are still giving them more content than we strictly ought to. For example, let us say 'the dog thinks there is a cat up the tree', rather than 'the dog thinks there is a fairly cat-like object in the tree', but let us not automatically infer the dog's belief includes the whole biological load which the word 'cat' might carry. (p. 117)

If we follow Bennett, we still seem to have the problem of identifying a particular proposition as the content of a neural state.

Moreover, there remains another crucial feature of intentional states, perhaps the most crucial feature, that neither this theory nor Dennett's accommodates. The crucial feature of intentionality still missing from these theories is intensionality.[14] Mental states are intensional while registrations are not. Intensionality is a feature that psychological states have and that biological states, events, and processes do not. Since registrations also lack this character, they have been naturalized *too far*.

The intensionality of psychological states consists *at least* in this: take any true attribution of the belief that object b has property F,

---

14. This lacuna is particularly ironical in the case of Dennett's approach, for his work is notable for its insistence on intensionality as a mark of the mental.

and substitute into the attribution terms coreferential with *b* and/or coextensive with *F*. Sometimes the result will be a false attribution of belief. For example, 'Oedipus believed that Jocasta was the queeen of Thebes.' But if we substitute for 'Jocasta' the coreferring term 'Oedipus' mother', we shall produce the presumably false statement that 'Oedipus believed his mother was queen of Thebes.' And if we substitute for 'queen of Thebes' the coextensive predicate 'Oedipus' bride', we will produce the equally false statement that 'Oedipus believed that his mother was his bride.' By contrast, such substitutions do not enable us to produce a false statement from a true one in biological, chemical, or physical theory. For example, such substitutions in attributions of fitness levels or adaptations may produce unlikely or perhaps even comic results, but the sentences so produced will for all their uselessness still be true.

What are we to make of this difference? Does it affect only the sentences that report these facts, or is it a symptom that reflects some difference between psychological and biological states? If it does reflect difference, then what are its ramifications for the general strategy of naturalizing intentional states or the specific attempt to certify intentional psychology as respectable on analogy with evolutionary theory?

We may quickly dispose of the notion that intensionality is merely a feature of sentences that is not symptomatic of any facts about the world. On this view, the attribution of intentional states to objects is a heuristic device, a calculational convenience we employ because an extensional approach to humans and other complex systems (animals, computers, extraterrestrials) will not enable creatures like us to complete and communicate real-time explanations and predictions of their behavior. As noted above, this approach would seriously compromise a comparison between fitness and intentionality. If fitness is a real property of organisms and their components, while intentionality is just a heuristic overlay on nonintensional phenomena, then the analogy breaks down seriously. There will be little reason to suppose that folk psychology treated as a heuristic device can provide the initial stepping stones to a scientific psychology. The analogy would hold if we took to treating fitness attributions as convenient fictions employed because nonevolutionary descriptions cannot be related in manageable ways to differential reproduction and evolution. But doing this is curing the disease while killing the patient, so far as the appeal of naturalism is concerned. In effect it denies the theory of natural selection its place as a scientific theory. It thereby jeopardizes the respectability of intentional psychology beyond salvation. No one seriously committed to an intentional research program in psychology can expect to treat the intensionality of psychological sentences as anything less than a reflection of the character of the states they advert to. And of course, no serious exponent of intentional psychology does. The intensionality of psychological states is asserted to reflect the fact that psychological states really do have content, that they are or contain representations of the way the world is or could be.

Dennett's and Bennett's naturalistic approach to intentional states does not reflect their intensionality. It does explain how they might have content, but not how this content can generate intensionality. The content that naturalism accords to intentional states is nothing more than the purely extensional content that evolutionary naturalism accords to fitness levels. Pursuing the evolutionary model has provided an explanation of adaptational content, but not of semantic content. So an intentional belief can hardly be identical to an extensional registration, not even a highly educable one.

Bennett's account of registrations reveals that they will not be intensional—that the substitution of coreferring terms and coextensive predicates in the proposition an organism is said to register cannot convert that report from true to false. Given the notion that belief is nothing but educable registration, Bennett argues that if P logically entails Q, then no organism without language can register P without registering Q: "If P entails Q, then any environment in which P is also one in which Q and there cannot be any saving differences of time or of sense-modality which would let us connect a with P but not with Q" (p. 114). Similarly for logically equivalent statements: "For any languageless animal a, it seems, if a thinks that P then its behavior cannot positively conflict with the attribution to it of the belief that Q where Q is logically equivalent to P" (p. 116). Bennett adds the qualification that considerations of parsimony can rule out the attribution of certain artificially constructed logical equivalents. Such hoked-up statements contain extraneous components that do affect truth value, but do not increase the explanatory power of the registration with respect to any behavior. Clearly, registrations are extensional for almost any logically equivalent propositions not hoked up by the logician as counterexamples to an extensional criterion of significance.

Although Dennett's upgrading account of intentional states does not proceed at the level of constructional detail that Bennett's does, it is just as firmly committed to the extensionality of the content it allows

us to attribute to neural states. This is because of its agreement with
Bennett on the role of behavior in shaping these contents.

> No afferent can be said to have the significance 'A' until it is
> 'taken' to have the significance 'A' by the efferent side of the
> brain, which means, unmetaphorically, until the efferent side of
> the brain has produced a response (or laid down response con-
> trols) the unimpeded function of which would be appropriate to
> having been stimulated by an A. This is not the epistemological
> point that as behaviorists *we cannot tell* whether the organism's
> brain has discriminated its stimulus as having the significance A
> until the organism manifests this in behavior, but the logical or
> conceptual point that *it makes no sense* to suppose that the dis-
> crimination of stimuli *by their significance* can occur solely on
> the afferent side of the brain. (p. 74)

If anything, Dennett is even more firmly committed to the extensionality of the contents of neural states, since for him the role of behavior
in discrimination of content is explicitly a matter of conceptual truth,
while Bennett is silent on the modality of his version of this claim.

The upshot is that the details of the evolutionary approach to the
naturalization of belief don't quite do the job. They come close; they
provide what might be necessary conditions of psychological content,
but not sufficient ones. The contents they accord to neural states and
registrations are not intensional. If the content of beliefs is intensional,
then registration is not yet belief.

FROM REGISTRATION TO INTENSIONALITY

What is missing from registration that prevents it from accounting for
intensionality? Can we find something that when added will bring with
it intensionality, while still preserving analogies to natural selection
crucial both for the naturalization of psychology and the respectability
of intentional theories of it?

To answer these questions, consider the causal explanation for a
failure of coreferring terms or coextensive predicates to preserve the
truth of a psychological attribution. Why does Oedipus believe that
Jocasta is the queen, but not that Jocasta is his mother, even though
'his mother' and 'Jocasta' corefer? Because he is ignorant of the facts
of coreference, or because he has erroneous beliefs about logically valid
inferences. Apparently, the intensionality of a sentence reporting one

psychological state is a function of the agent's other intentional states, and the lack of others. Consider a cognitive agent who is omniscient, both about logical matters and contingent ones. Would the omniscient agent's psychological states reveal their intensionality in the substitution of terms and predicates? No. The omniscient agent would know all the names and descriptions under which all objects fell, and would never make a false inference or miss a logical implication. We are not omniscient, and the intensionality of our intentional states shows the dependence of their identities on what else we believe and desire.

That a particular mental state has some content or other may not depend on the contents of the other neural states collateral with it. If naturalism is correct, it may depend only on its environmental appropriateness understood as the fortuitous product of variation and selection. But exactly what its content is, what statement in what vocabulary most accurately captures its content, is a causal function of the contents of the other psychological states obtaining in the system along with it.[15] So our problem in identifying the exact content of a psychological state is ignorance of the vast and diverse class of other contentful states cooccurring with the intentional state we want to identify. If we could identify all the psychological states that work together to produce the environmentally significant behavior the organism emits, then English might suffice to describe them all and to specify even a languageless creature's psychological states with all the precision we desire. To do this, English need only be rich enough to describe all the organism's potential environmental stimuli and its behavioral repertoire with accuracy and detail: the correct description of a large enough body of behavior in a large enough environment is sufficient to describe content, on the naturalistic approach. We may hope eventually to extend this method of specifying content to creatures with language. Indeed, it must eventually provide a more accurate means of identifying psychological content than our everyday reliance on verbal behavior provides. Otherwise naturalism will be no improvement on folk psychology. Of course, the instrumentalist interpretation of the theory is in effect the claim that there is no such improvement to be expected.

---

15. Davidson (*Essays on Actions and Events,* p. 68) expresses this point forcefully: "A belief is identified by its location in a pattern of beliefs; it is this pattern that determines the subject-matter of the belief, what the belief is about. Before some object in, or aspect of the world, can become part of the subject-matter of a belief (true or false) there must be endless true beliefs about the subject matter." Presumably "endless" is not to be taken literally.

Thus, when coreferring or coextensive terms are substituted into a true psychological statement, the truth or falsity of the resulting statements is a function of three different possibilities.

1. Preservation of identity: sometimes, if a believes that Fb, and if (x) Fx IFF Gx, or $b = c$, then the substitution of G for F or c for b in the statement that a believes that Fb preserves truth because 'a believes Gb' or 'a believes Fc' or 'a believes Gc' describes the very same belief. Thus my belief that Paris is the capital of France is the same belief as my belief that the City of Lights is the capital of France, and the same belief as my belief that the City of Lights is the capital of the hexagon. For others with different beliefs, differently organized, these substitutions will not preserve the identity of the belief referred to.

2. Preservation of truth value: sometimes the same substitutions will not preserve the reference of the statement to b, but will preserve its truth under substitution, because the subject does, as a matter of fact, have a second, different belief about c, which satisfies the statement under its substitutions. Thus my belief that nine is the number of the planets is a different belief than my belief that nine is the number of fielders in baseball, but substitution of the predicate 'is the number of the planets' for 'is the number of fielders in baseball' preserves the truth of a belief attribution, because I have the relevant further belief.

3. Conversion to falsity: often, however, substitutions change truth values of the resulting statements because a simply does not have the belief that Fb or Gc, even when $(x)Fx$ IFF Gx and $b = c$. Oedipus' beliefs about Jocasta, his mother, and the monarchy of Thebes provide vivid examples.

Which one of these three alternatives obtains is a function of what other beliefs the subject has and how they are organized. The beliefs of an omniscient agent will fall exclusively under cases 1 and 2 above. For an organism with only one belief, substitutions will fall only under cases 1 and 3. Between these two poles lie the rest of us. Thus the intensionality of beliefs is a function of the presence of other beliefs, because the content of a belief is a function of other beliefs.

Now, registrations and the contents of neural states are not like this: they are extensional. The substitution of coreferring terms and coextensive predicates in them will never provide grounds to suppose that cases 2 and 3 obtain, because their content is a function only of the appropriateness of outputs given environments, and not of other registrations and neural contents (except where their outputs are other registrations or contentful neural processes). This suggests a natural addition to theories of registration and neural content that may generate intensionality.

According to Bennett an educable registration is a belief. Since registrations, educable or not, are not intensional, this cannot be correct. But what if a single belief is a *set* of educable registrations? Similarly, a belief may be construed on Dennett's terms as a set of very many neural states, the sum of whose contents give the content of the belief. For example, consider a male polar bear who comes to have an intentional state that we would perhaps parochially describe as the belief that a seal is in the vicinity. He will be in this state because he registers many propositions. Which ones? Well, we cannot identify any of them until we know a good deal about the polar bear's neurological organization. And the propositions that it registers will be quite unlike those the ethologist might attribute to it as "beliefs" on the basis of its behavior. For example, the bear's registrations may include the propositions that acoustic vibrations characteristics of Phocidae mating behavior are being propagated along a certain acoustic gradient (i.e., the seal is flapping the ice with its flippers), that a surface is reflecting light of 630 nanometers' wavelength from location x, y, z (i.e., the seal is brown), that Phocidae pheromones are present in concentrations above the bear's olfactory threshold with a directional gradient (i.e., the bear smells the seal's odor coming from a certain direction), that the olfactory gradient matches the acoustic one, and so forth. For all we now know, none of these examples are the correct attributions of a polar bear's registrational contents. Doubtless, science will reveal exactly what the bear registers, and when the full extensional story about its sensory and cognitive apparatus is known, we will be able to give a complete list of the bear's registrations on this occasion.

Note the bear believes none of these propositions it registers. After all, what does it know of acoustic and olfactory gradients, nanometrically measured wavelengths of light, or the correct Linnaean classification of its prey? But it registers all of them and their logical equivalents (for registrations are extensional). Depending on its behavior, we may call this set of registrations the belief that a seal is present at a given location. Or we may decide not to call it a belief, if our favored theory does not involve attributing beliefs to bears or an ontology of seals to their belief systems. Regardless, the propositions will still accurately reflect his registrations no matter what the bear's unvoiced linguistic abilities or conceptual scheme.

The hypothesis that every particular occurrent belief token is *identi-*

cal to a set of registrations must be distinguished from the weaker thesis that beliefs are distinct states of an organism somehow caused by neural registrations. We already know that intensional effects, psychological states, are the result of nonintensional causes; even dualists accept this view. What we want from naturalism is a theory about exactly how intensional states come about. So we need something stronger than a cause-and-effect story. An identity theory, that belief B equals registrations $\{r_1, \ldots, r_n\}$, provides such an explanation.

The identity thesis is not a claim about types of beliefs and types of sets of registrations. It is not that all members of the class of everyone's beliefs that Paris is the capital of France are constituted by members of a single set of registrations. The identity hypothesized is a token thesis about the identity between an agent's particular occurrent belief state and a particular set of registrations. Assuming that the intentional is at most supervenient on the nonintentional, as naturalists do, we should expect that two different instances or tokens of the same type of belief will be identical to particular sets of registrations of different types.

This identification of belief tokens with sets of registration tokens can produce intensionality quite simply. Suppose that a's belief that Fb is identical to a particular set of purely nonintensional registrations, a fairly large set. a's belief that Fb will be identical to a's belief that Fc, or for that matter a's belief that Gc, if the set of registrations that constitute either of these latter beliefs is a subset of the set of registrations that constitute a's belief that Fb. In this case substitution preserves the reference of the original belief ascription.

Suppose that the extensions of the sets of registrations for the three belief states Fb, Fc, and Gc diverge. Then substitution of coreferring terms or coextensive predicates in the original description of a's belief that Fb will preserve truth just in case a, also affects all the registrations in the set that constitutes the belief that Gb or Gc. These sets will doubtless overlap with the set of registrations that constitute a's belief that Fb, but their diversity from this set is what makes the attribution to a of the belief that Fb or Gc the attribution of a different belief. Of course, when a does not affect all the registrations required for the belief that Fb or Gc, it will simply be false that a has either of these beliefs. Voilà intensionality.

Defenders of intentional psychology can take some satisfaction that this strategy looks rather like that of "stupid homuncularism." Homuncularism in biology is the thesis that the fetus develops into the child's body because it already is one, though too small to be recognized. Homuncularism in biology is unsatisfactory, for it simply pushes back the question of how the fetus develops into the baby's body into the question of how the egg and sperm develop into the fetus's body. In psychology, homuncularism is the name for a similar fallacy, that of explaining the intentionality of thought by appeal to a thinking subsystem of the brain, which has access to the rest of the brain's information input, storage, and symbol manipulation system. Homuncularism simply shifts the question of how the brain thinks back to the question of how this subsystem thinks.

As Dennett points out, "Homunculi are *bogeymen* only if they duplicate *entire* the talents they are rung in to explain. . . . If one can get a team or committee of relatively ignorant, narrow-minded, blind homunculi to produce intelligent behavior of the whole, this is progress."[16] These ignorant, narrow-minded subsystems of a intentional system are the "stupid homunculi." Now the present proposal takes "stupid homuncularism" one better. While the homunculi Dennett contemplates are still intentional, registrations are so narrow-minded that they are no longer intensional, and therefore not intentional either. In effect, they are the end points of the research strategy Dennett proposes: "Eventually this nesting of boxes within boxes lands you with homunculi so stupid . . . that they can be, as one says, 'replaced by a machine.' One *discharges* fancy homunculi from one's scheme by organizing armies of such idiots to do the work" (p. 124). It is compatible with the notion that intentional states are sets of registrations or sets of neural events with nonintensional content that there be "levels of organization" between beliefs on the top, so to speak, and registrations on the bottom. Within the set that constitutes a given belief, say, there may be natural divisions into subsets of representations, common to other beliefs, and other intentional states, which work together or which allow the beliefs they compose to work together to produce other psychological states and behavior. Theories about such subsets consist in generalizations about intentional subsystems simpler than the variables of folk psychology, subsystems that underlie the operation of or compose such variables. They are Dennett's stupid homunculi, and they are the stuff of the intentional psychological theories that succeed and improve upon folk psychology. The intentionality of such subsystems consists in the same features that make for intensionality of belief,

16. *Brainstorms*, pp. 123–124. Pages in text refer to this work.

desire, and so forth. The sets of registrations they consist in, though smaller than what qualify as beliefs, for instance, are still large, and are identical, intersecting, or exclusive. Registrations turn out to be no more tractable than Mendelian genes, and probably less so. Even deprived of their intentionality, mental state descriptions like this are still heuristic devices for cognitive agents like us.

Types, Tokens, and Interesting Generalizations

Consider the thesis that beliefs are sets of registrations. This claim is ambiguous as between types and tokens. It could mean that every particular dated occurrent belief of some particular agent is identical with a set of particular registrations or contentful neural events realized in the agent's neural apparatus on that occasion. This is the token identity of beliefs and registrations.

Then there is type identity, the identity of kinds of intentional states with kinds of sets of registrations. These type identities could come in varying degrees of strength.

1. It could be that every belief that P realized by any cognitive agent is the realization of exactly the same (type of) set of registrations.
2. More weakly, it could be that each and every instance of a belief that P is the realization of one member of a disjunction of sets of registrations, a small or a very large disjunction.
3. If the disjunction is very large, it might still be the case that, for some narrower class, species, or population of agents, the belief that P is always realized by a single or a small number of type of set of registrations.
4. If the disjunction is very large, it might be that, if I believe that P, then if anyone else has exactly my registrations, he or she will also believe that P, but not vice versa. This alternative expresses the supervenience of intentional states on registrations. Here the class of agents in which the type for the notion of supervenience identity holds is the narrowest possible.
5. It may be that on different occasions, when I have numerically distinct beliefs that P (as when I have an indexical belief or when I first believe P, then disbelieve it, and come to believe it again, on the basis of new evidence), each realization by each individual of the belief that P consists in a qualitatively different set of registrations.

Which of these type identities between intentional states and nonintentional ones can we expect to obtain? In part this depends on what proposition we substitute for *P*, or so it appears. Intuitively, one would suppose that the "simpler" and more common the belief, the closer its content is to that of an observational report, the more likely it is to be type identical over many agents with the same relatively small disjunction of representations.

Consider the intentional states studied in psychophysics: an experimental subject's belief that he has just seen a flash of light, or that one noise is louder than another by a just noticeable difference. These kinds of beliefs can plausibly be identical to a small number of registrations, restricted to one sensory modality and its neural projections. We may expect that this small set of registrations is common to many members of a population or species because they are similarly equipped by way of sensory organs, and so forth. Our evidence for such identities will be the similarity of controlled responses by subjects in contrived settings to quantitatively determinable environmental changes. More complex intentional states, for example beliefs about a particular material object, are based on more than one sensory modality. Each instance of such a belief is identical with a larger set of registrations, and two subjects may have the same type of belief about the same material object, while the registrations that constitute these beliefs diverge, for reasons of physiology, location, and so forth.

The simpler a belief, that is, the smaller the number of registrations to which it is type identical, the easier it will be to uncover interesting psychological generalizations about it. Consider the generalization that any agent who believes that *P* and believes that if *P* then *Q*, also believes that *Q*. For most people, and for most propositions, this generalization is obviously false. For some people and for some propositions, however, it seems more reasonable. Why? Not because there is a class of people with greater than average logical acumen, but rather because some beliefs "go together" more frequently than others: these will be relatively simple beliefs, like the perceptual ones discussed above. Beliefs simple enough to go together in such a way as to regularly instantiate the *modus ponens* generalization above will do so presumably because of relations between the sets of registrations that compose them. In the simplest case, a triple of belief types that always behave in accordance with this generalization do so because the registrations that realize the first two beliefs either include the ones that realize the last, or else are nomologically cooccurrent with them. If we hold that beliefs are connected causally, in laws that we can discover and employ, then a commitment to the mereological uniformity of nature leads to the conclusion that there are manageable type identities between beliefs and the nonintentional registrations that realize them. The smaller the sets of registrations that beliefs can consist in, the easier these laws and the type identities they require will be to discover.

The simplest beliefs, the ones studied in psychophysics, will thus be the most tractable. The regularities describing their relations to other beliefs, to the environment, and to behavior, will be the easiest to uncover and make use of. Perceptual beliefs are the smallest sets of constitutive registrations and have the most stereotyped efferents and the most unambiguous environmental causes in stimulation. The more complex the beliefs studied above this level, the more difficult it will be to uncover useful generalizations. If interesting generalizations are not uncoverable about beliefs at the level of psychophysics, then the underlying registrations that realize a given type of sensory belief are not sufficiently homogeneous to be the subjects of laws. They do not have a small set of homogeneous effects the psychophysicist can conveniently measure. In this case psychophysics would have to turn to the level of registrations themselves to find interesting generalizations. That is, it would have to turn to nonintentional characterizations of the neurophysiological processes of perception. And if psychophysics must turn to registrations to uncover its laws, then so will the rest of psychology. To do so, however, is to give up the hope for an intentional science, for registrations are not intensional.

What is meant here by "interesting generalizations"? Since the generalizations of psychology can be no stronger than those of biology, what is not meant are impeccably exceptionless universal laws about widely instantiated natural kinds. Like the rest of functional biology, psychological theory will be restricted to uncovering the mechanisms of "model systems"—the cat's optical channel, or the frog's fly-catching technique, or human memory storage. The most we can hope for is to generate hypotheses useful to agents like us, applicable (with exceptions we cannot systematically explain) at most to neighboring systems of similar structure and capacity. Generalizations with both much content and few exceptions will be limited, in biology and in psychology, to particular species or to smaller sets of biologically homogeneous systems. As with biological claims, the more restricted the domain the less systematic will be the theoretical ramifications of these generalizations for other systems, or even the same systems under different conditions.

The less restricted the domain of such generalizations, the more likely they are to be false, or if not, then they will be hedged around with so many qualifications and expressed so generically, that they will have little predictive content with respect to any variable we can effectively measure.

Prima facie, the richer the content of an intentional state, the more disjunctive the sets of registrations to which it will be type identical, and the more difficult it will be to express and employ the generalizations in which it might figure, even on the understanding that such generalizations are restricted to one or a small number of model systems. Accordingly, the prospects are low for any really useful intentional generalizations with considerable content. One way to deal with this problem is to decompose rich intentional states into sets of poorer ones, intentional states of stupid homunculi; then one can search for interesting generalizations at this lower level of organization. The result will be a theory, not about intentional states ordinarily understood, but about theoretical entities to which these states are related. But the relation between personal states and these stupid homuncular ones will be that of token-token identity and not type-type identity. If there are type-type identities, then the stupid homuncular states and the smart personal ones, like desires and beliefs, will stand or fall together.

What if the only interesting generalizations in psychology more useful than those known to folk psychology are at the level of registrations alone? What if there are no humanly manageable kinds above the level of the single or small number of registrations, no package of registrations that are more regularly produced by the same environments and that result more reliably in the same single or small number of efferent effects than the exception-ridden generalizations of folk psychology? This prospect would mean nothing less than the impossibility of an intentional psychology even as useful as the claims of biological science.

CHAPTER EIGHT

Biology and the Behavioral Sciences

N OW WE CAN FINALLY see on what the analogical appeal of intentional psychology to evolutionary biology turns. For deciding whether there is a strong enough analogy to pursue intentional psychology in the face of its weaknesses, the crucial question is, can we expect intentional psychological theory to be anywhere near as heuristically powerful for cognitive agents of our powers as biology is? (Or if one thinks intentional psychology is already as or more powerful, can we exepct it to increase in heuristic power for agents like us in the way that nonmolecular biology has?) The answer here turns on nothing other than the question of whether the kind terms of intentional psychology are more heterogeneous in their relations to the terms of biology than those terms are heterogeneous in their relation to the kind terms of physics and chemistry.

ELIMINATIVISM AND INTENTIONAL PSYCHOLOGY

The degree of explanatory and predictive success biology has achieved turns on the number of disjuncts in the type identities for biological kinds. The same is true for intentional psychology. Evolutionary biology is both predictively weak and theoretically autonomous because the type identities between its variables and those of other theories are too complex to permit anything like the reduction of this theory to nonevolutionary ones. Variation and selection are frequent and stable enough phenomena for "interesting" evolutionary generalizations to have been uncovered. Heredity too has revealed the operation of "laws" simple enough and precise enough to enable cognitive agents like us to explain and predict a variety of genetic phenomena. But we now know why the generalizations of evolutionary biology have remained unimprovably imprecise: the kinds of entities and processes these statements mention are not identical to small and manageable

169

classes of kinds described in the generalizations of nonbiological theories. The type to which they are identical are complex, disjunctive, and still largely undiscovered.

For all their complexity, these type identities are just manageable enough to permit something approaching reduction to molecular mechanisms of some species-specific instances of evolutionary and genetic regularities, or instances of their operation in well-understood "model systems." These restricted generalizations are reduced to regularities in molecular biology, for example, through the provision of type identities between functional or genetic predicates and biochemical ones. This is how evolutionary biology links up to the rest of science. It is beyond argument that the type identity between the variables of intentional psychology and nonintentional neuroscience are also too complex to permit the sort of smooth "layer-cake" reduction familiar from physics. To this extent the analogy with evolutionary theory's typology is unquestioned. The question thus becomes whether, like the types of evolutionary biology, intentional kinds are simple enough to permit even case-by-case reduction, along the lines of what Philip Kitcher calls "explanatory extension." That is, the question is whether interesting generalizations of intentional psychology about model systems exist that can be linked to neuroscience, in the way evolutionary generalizations can be linked to molecular biology.

Few will assert that such interesting generalizations are already in hand. Intentional psychology has produced nothing to rival even Mendel's primitive hypotheses about genetic control of the phenotype, let alone their less exception-ridden and more precise successors. Still less plausible is the suggestion that such interesting generalizations are already embedded in the folk psychology we employ everyday.

The problem of defending intentional psychology begins with the recognition that any such implicit laws are either patently false or next to definitional. No defense of intentional psychology would be necessary if it were granted that such laws already exist. In their absence the only assurance that improvable generalizations of an intentional psychology are yet to be discovered is the existence of type-type identities at least simple enough to allow for case-by-case reduction. The possibility of such a unification with neuroscience is at least a necessary condition of scientific fruitfulness, particularly for increased predictive precision and range.

Since registrations are nonintentional, to vindicate intentional psychology what we will need is more than the existence of sufficiently simple type-type identities between types of individual registrations and species-specific neurological states. Such type-type identities between registrations and neural states are necessary, and it is by no means clear that our neurology is simple enough to provide them. But in addition, we will need type-type identities between neural states and *sets* of registrations that are at least large enough to make for the intensionality of the beliefs they constitute.

How simple will these type-type identities have to be for case-by-case reduction? Consider a biological example, one touched on already and on which philosophers have dwelt almost since their earliest examination of contemporary biological theory: transmission genetics explains why some children of phenotypically normal heterozygous parents are afflicted with sickle-cell anemia. The gene for this phenotype is recessive and appears in offspring only when both parents provide a gene for this phenotype, which results in a homozygous child showing the phenotypic anemia. Before we consider how this explanation is grounded in molecular biology, there are some things worth noting about it. One is that, independent of any further biochemical information, biologists already have good grounds to believe the "laws" of transmission genetics are interesting generalizations. That is, biologists have good reason to believe that terms like 'recessive', 'homozygous', 'dominant', 'heterozygous' pick out something like natural kinds, even though they are restricted to eukaryotic species, and even when we knew nothing about the disjunctions of nongenetic kinds with which they are identical. In particular, biologists had already identified a trait as phenotypic, because it assorts in Mendelian ratios.

Now, as Patricia Kitcher notes, "though we have answered the original question, the answer raises further questions, for example, Why do heterozygotes avoid the disease, while homozygotes are invariably afflicted?"[1] The answer is that the sickle-cell gene produces an abnormal hemoglobin protein whose molecular structure causes the blood cells, which are composed of hemoglobin, to sickle and renders them incapable of transporting oxygen. When the sickle cell-gene is accompanied by a normal one, in the heterozygote, enough normal hemoglobin is produced to prevent the sickling of blood cells. One of the great triumphs of modern molecular biology is to have (1) isolated the *single* exact amino-acid difference from normal molecular composition that is common and peculiar to all sickle-cell hemoglobins; (2) identified

---

1. Kitcher, "In Defense of Intentional Psychology," pp. 102–104.

the single difference in the polynucleotide sequence of the DNA of the sickle-cell gene that causes the difference in protein structure; and (3) determined the *single* exact allosteric process by which it causes sickling in all cases. Biochemistry's success in solving this incredibly complicated problem is due to the fact, that so far as molecular biology is concerned, it is *simple:* there is just one amino-acid difference, just one molecular mechanism with the sickling effect, just one difference in the genetic code that produces it, and lots of hemoglobin to experiment on.

What exactly is the relevance of this work to intentional psychology? Kitcher claims we should expect the same process of scientific advance in psychology.

> It seems obvious that some more basic sciences have deepened the understanding of the inheritance of sickle-cell anemia originally provided by transmission genetics. A rough characterization is that they have deepened that explanation by *extending* our understanding of some of the facts cited in the original explanation. Explanatory extension is not reduction. In this case there is no prospect of recasting the original explanation in chemical terms. How could one formulate the crucial premise, that the normal parents produce abnormal offspring because they reproduce sexually, in the terminology of chemistry?
>
> . . . The model of explanatory extension [case-by-case reduction] just illustrated seems directly applicable to intentional psychology, with its layers of functional decomposition leading ultimately to the neurophysiological level. In fact, if the problem of intentional psychology succeeds, then the relation of explanation extension will hold between psychology and neurophysiology (and chemistry and physics). . . . So although intentional psychology is irreducible to a more basic science, it would be fully integrated into the scientific picture of the world. (p. 104)

As a claim about logical possibilities, this one is unimpeachable. But as a claim about the foreseeable prospects of intentional psychology, it is wildly optimistic. Explanatory extension of the kind that underwrites transmission genetics is simply not in the cards for intentional psychology. Its differences from the evolutionary case are just too great.

Note first that, though there is no reduction in the genetic case, a number of its necessary conditions are present: there are several interesting generalizations about the transmission of phenotypes and genotypes and their contributions to fitness; and there are some relatively manageable type-type intertheoretical identities. These identities are necessary for this explanatory extension of transmission genetics. In particular, the sickle-cell syndrome is type identical to a single physiological phenomenon, the deformation of red blood cells; the sickle-cell protein is type identical to a small disjunction of primary sequences of amino acids; and the sickle-cell gene's recessive mutation is type identical to a single difference in the polynucleotide sequence of the hemoglobin gene. These type identities are crucial to the very possibility of an explanatory extension of the particular transmission genetic phenomenon. *More important,* they are crucial to the existence of interesting evolutionary generalizations about the sickle-cell trait. If the type identities in question had not been so simple, there would have been no regularity about the balanced polymorphism of the sickle-cell trait to extend the explanation of.

Discovering interesting genetic transmission "laws" was the first step in this explanatory extension. These laws would not have been discovered had the type identities not been strong enough to permit their explanatory extension. It is patent that we have not discovered anything like the transmission "laws" in intentional psychology, heuristically interesting generalizations that cry out for reduction by cases. The explanation is this: though there are interesting generalizations at the level of neurophysiology, the type identities between intentional and neural states are too complex to generate even useful regularities at the intentional level, beyond the venerable ones of folk psychology. Recall that we need two levels of type identity here: identities between registrations and neural states that permit interesting generalizations about registrations and their causes and effects, and identities between registrations and intentional states that permit interesting generalizations about intentional states and their causes and effects.

What we know now about the brain suggests that the neural states whose registrations would constitute a given intentional state are huge in number, diverse in molecular structure, various in their interconnections, and widespread over the cerebrum, cerebellum, and medulla. For example, my belief that there is a tomato in front of me right now involves registrations in the optic center at the rear of the brain, registrations at my olfactory centers at its side, registrations in my memory centers for the notion of 'tomato', and a great deal more besides. And the next time I come to believe there is a tomato in front of me, the set of registrations involved will be different. What is more, the registration of statements about the visual field is a disparate and distributed process.

As Churchland reports in his introduction to neuroscience for philosophy,

> Recent studies indicate that distinct topographical maps of the retina are scattered throughout the cortical surface, and enjoy distinct projections from the lateral geniculate or from elsewhere in the thalamus. The hierarchical system of topographic maps which culminates in the "secondary visual cortex" at the rear of the brain is thus only one of several parallel systems, each processing different aspects of visual input. . . . Similar complexities attend the "somato-sensory cortex" which emerges as only one of several parallel systems processing different types of somato-sensory information: light touch, deep pressure, limb position pain, temperature.[2]

There is nothing like this sort of complexity in the sickle-cell hemoglobin case. The sickle-cell hemoglobin story was relatively simple to uncover, and most explanatory extensions of evolutionary phenomena will doubtless be far more difficult. The disanalogy with intentional psychology is that the simplest case-by-case reduction of any intentional phenomena will be vastly more complex than the hemoglobin case. The complexity neuroscience is revealing, together with the absence of interesting intentional generalizations after millennia of searching, is decisive evidence for this conclusion.

If intentional psychology hopes to provide a reduction of the causes and consequences of the typical thought that flits through a person's mind on the model of evolutionary biology's explanation of the causes and consequences of an organism's or a gene's adaptation, it has an impossible job before it. The difficulties are negligibly less daunting for the problem of extending the stereotypical process of thinking through a chess move or proving a theorem in geometry. First of all, we have in these cases nothing like a phenomenon as stable as the phenotypic transmission that gave the case-by-case reduction of evolutionary regularities their start. We have generalizations, it is true, extractable from folk psychology, but they do not compare for predictive content and precision with even so lamentably false a generalization as, for instance, the Mendelian law of independent assortment.

---

2. Paul Churchland, *Matter and Mind* (Cambridge: MIT Press, 1984), p. 139.

At best the generalizations of folk psychology have the standing of the principle of natural selection *before* biologists knew anything about hereditary transmission. Before biologists knew this, the claim that the fittest survive had limited predictive content, for the only measure of fitness was survival. Without identified registrations the generalizations of folk psychology are in the same boat. Before the physical localization of the genetic material, the claim that genes assort independently had little predictive content, for our only measure of independent assortment was the distribution of the phenotypes that independent assortment of genes was to explain. Similarly, the claim that registrations go together in ways that explain intentional phenomena has little content unless we can locate registrations in the neural material that constitutes them. And the intensionality of beliefs, for example, shows that the number of registrations we need to identify to extend intentional explanations will be huge. By contrast, the number of proteins and genes we needed to identify to extend the explanation of sickle-cell anemia's transmission was minimal. The number of neural states we would need to identify to extend the explanation of any regularity about registrations will be huge. Molecular biologists were able to extend the genetic explanation of sickle-cell anemia to the molecular level, because there was only one molecular arrangement to which the defective hemoglobin gene is identical.

The conclusion is not the logical impossibility of a neuroscientific reduction of any type of intentional state. Rather, it would be orders of magnitude more difficult than the evolutionary case, and utterly without scientific utility. It would be too complex for real-time prediction or unifying explanation by agents of our cognitive power. But if it is without scientific utility, then intentional psychology is impossible, not logically, but as a practically significant science. The possibility of explanatory extension is so weak a requirement on science that any theory that fails it can have little to say for itself.

Someone might admit that an intentional psychology of everyday life is beyond us but that this should not be the aim of intentional psychology. Rather, it should aim at the intentional systematization of stereotyped phenomena like psychophysics, chess playing, laboratory memory and recognition tasks, image rotation, and so forth, much as evolutionary biology focuses on its manageable problems, even when they are somewhat artificial or otherwise special. It may be just possible to generate a psychology of registrations for such stereotyped phenomena, a set of interesting generalizations about registrations, their environmental stimuli, and behavioral consequences. This would be something of a vindication of naturalism—but not a vindication of a naturalistic intentional psychology. Registrations are just not intentional. If we had such a nonintentional theory of the registrational causes of laboratory behavior, we would not be able to usefully employ it for real-time prediction outside the laboratory, because of the diversity of neural phenomena that disjunctively realize any very interesting registrations.

It is ironic that an exploration of features that make the analogy between evolutionary biology and intentional psychology attractive leads to a pessimistic conclusion about the latter's prospects. To the degree that intentional psychology and its explanatory variables mirror features of evolutionary biology, their weaknesses are more fully revealed. Naturalism is correct: intentional states have their content as a result of selection on variation, just as adaptations do. And at least some types of intensionality are the result of mechanisms on which this shaping process operates. Moreover, intentional psychology is irreducible to the rest of science because, like evolutionary theory, its variables supervene on those of the rest of science. Differences in the degree of complexity in this supervenience make intentional psychology impossible for cognitive agents like us, while biological theory remains not only possible but necessary for our survival.

If naturalism is the only basis on which intentional psychology can be legitimated, then as a body of interesting generalizations and theories that transcend folk psychology, intentional psychology is not after all a science worth pursuing. At most, naturalism shows how intentional psychology can be possible as a natural disposition of *Homo sapiens*. It shows how we might naturally have come to adopt and employ folk psychology in everyday life, because for that purpose it is the only practically useful instrument. Naturalism is as good an explanation for the origin and persistence of folk psychology as we can expect. But it is no scientific vindication of it or its intentional successors.

Intentional psychology's problems stem from its being a discipline twice over supervenient on discoverable regularities and laws. It is unlikely to produce any results as impressive as those produced in biological science. On the other hand, a nonintentional psychology can expect to be no worse than biology in uncovering results of enduring scientific interest, for nonintentional psychology just in biology.

Results of an Instrumental Science

What sorts of results can we expect in an instrumental science like biology, and how interesting will they be? The answer is, all those achievements the history of this subject reveals to us: case studies, rough rule-of-thumb generalizations, quantitative models, underlying mechanisms for particular cases, natural history.

There will be case studies of species via their specimens, though, as genetics teaches us, the specimen is not the norm but at best an average around which much variation is normal. From these case studies we may infer useful generalizations, which we can further justify by bringing them under theory drawn from several disciplines. For example, the well-known generalization that, ceteris paribus, arctic species have a higher volume-to-surface-area ratio than nonarctic members of the same genus is an induction from ethology, underwritten by a combination of adaptational analysis and thermodynamics. But we cannot expect to upgrade this generalization into a law, for we cannot systematically reduce the scope of the ceteris paribus clause. We cannot determine in advance where the generalization fails, though when it does, we can identify the cause in the particular case. The reason we cannot upgrade our exception-ridden generalizations into laws is that the concepts we find convenient to express them, and to identify and exclude exceptions, are not themselves natural kinds, and the number of exceptions within our powers to identify are smaller than the number that obtain. This does not detract from the usefulness of biological generalizations for our purposes, only from their long-term prospects of incorporation and systematization in a general theory of increasing explanatory and predictive power.

We can expect quantitative, and particularly probabilistic, models that significantly increase our predictive powers with respect to aggregate outcomes, as we add variables and employ computational powers hitherto unavailable. This will happen not only in population genetics, as exponents of the semantic approach have reported, but also in molecular genetics. For example, having begun with a one-parameter model for random equiprobable nucleotide substitution in DNA sequences over evolutionary time (the Jukes-Kantor model), we may incorporate the knowledge that substitution is more frequent between purines and between pyrimidines than it is from purines to pyrimidines or vice versa. The result is a two-parameter model (the Kimura model) that enables us to provide an expected value for the epistemic probability of nucleotide divergences between lineages and species. We can confirm our expectations by the use of restriction endonucleases, DNA hybridization, and other techniques. These models applied to data provide the basis for molecular clocks that enable us to independently test paleontological dating of evolutionary events. But for our purposes it is crucial to keep in mind that the probabilities these models trade in, as with transmission genetics, remain resolutely epistemic. They reflect no indeterminism beyond the quantum mechanical causes of point mutation, and accordingly these probabilities are substitutes for nomological regularities beyond our cognitive capacities and also with little value, given our practical and technological needs.

Increasingly, natural history can expect to include more molecular detail, and to identify more causal chains and networks that have eventuated in the diversity observed on the earth. Some of this natural history will be the source of useful generalizations, as well as providing data that partially confirm models drawn from elsewhere.

Of course, biological research provides an account of the diverse mechanisms that realize effects, which we bring together under functions that are homogeneous with respect to our perspective on them. Perhaps the most striking advances in this strategy are to be found in developmental entomology. By the early 1980s, there were speculations about genes and proteins responsible for embryological change in a variety of insects. Terms like "diffusible morphogen," "cellular polarity," and "competition" described in functional terms molecules and processes for which there was no independent evidence, but which were hypothesized to explain how it is that the embryo makes the insect. Ten years later much of the evidence is in for the existence, localization, and mechanism of action of the genes that give substance to these explanatory concepts, at least in the fruit fly.[3] What is striking about the discovery of these mechanisms is that, even in so restricted a domain as the developmental sequence of *Drosophila melanogaster*, a domain presumably subject to the narrowest of constraints,[4] the developmental process is highly disjunctive and heavily interactive. The simplest regulatory gene for a morphogen, and the one molecular entomologists know the most about, the so-called *bicoid* gene, probably works by

3. For an introduction, with some historical anecdotes of interest to the historian and philosopher of biology, see Peter A. Lawrence, *The Making of a Fly* (Cambridge, MA: Blackwell Scientific, 1992).

4. On the narrowness of developmental constraints, see Lewontin and Gould, "Spandrels of San Marco and the Panglossian Paradigm."

producing a protein that binds to the DNA at the cite of another developmental gene, *hunchback,* which is itself characterized by a disjunction of different nucleic acid sequences. What is ultimately learned about the molecular biology of the development of the fruit fly will have significant application throughout developmental biology. But its influence will not be felt in general laws, or even very widely adaptable models. Rather, its influence will emerge in techniques for uncovering mechanisms, suggestions about alternative mechanisms to explore, and how to look for exceptions that undercut the applicability of models.

In all of the different compartments and subdisciplines of biology, research is heavily determined by technological interests, practical needs, and the desire to control nature up to limits we can easily discriminate by means effectively at our disposal. Even the most theoretical of biological scientists, who believe themselves to be searching for purely general regularities without any immediate instrumental payoff, will be willy-nilly engaged in an instrumental discipline. The concepts they employ will not be natural kinds, but will reflect the conceptual and cognitive powers that *Homo sapiens* have brought to their interaction with the biosphere since their first appearance on this planet.

Morals for Social Science

Most obviously, the prospects for sociobiology and behavioral ecology must be understood against the background of biology as an instrumental science. To begin with, we can no more expect an evolutionary theory of intentional traits, or rather, human behavior intentionally described, than we can expect interesting generalizations of an intentional psychology.

The instrumental character of biology should both reduce the pretensions of behavioral biologists and sociobiologists and blunt much of the criticism of their findings. On the one hand, researchers cannot expect models drawn from the study of other species to be vindicated in human behavior, not just because it is the result of cultural as well as genetic transmission, but because models don't generalize in an instrumental science. At the same time, opponents of controversial studies in behavioral genetics, for example studies of criminality, intelligence, schizophrenia, or depression, frequently condemn the entire enterprise as a disguised form of eugenics. They do so because hypotheses framed in this area are subject to exceptions and counterexamples, and are often withdrawn because they do not meet improving standards of evidence. Accordingly, in the absence of highly reliable results, opponents seek ideological motivation to explain the research interests of scientists working in this area.[5] But lack of universality, the need for ceteris paribus qualifications, the existence of exceptions, counterexamples, and the discovery of new evidence that undercuts models' applications, is characteristic of biological science. Results, no matter how well established, are likely to be so qualified and equally unlikely to sustain strong generalizations with significant predictive power. Biology is an instrumental science conditioned as much by its usefulness to us as by the way the world is arranged. Probably our practical concerns and interests in dealing with people are sufficiently different from our interests in other fauna and all flora, that results of the sort biology can offer are not reliable for practical intervention in this area. At best we will have models with the applicability of the most complex sort that population biology provides.

As for the other less self-consciously biological inquiries among the social sciences, biological instrumentalism translates into even less reliable, less replicable studies with even lower levels of generality and predictive power. These disciplines are irretrievably intentional, both in their accounts of individual behavior and in the terms under which they aggregate it into data about institutions and other social wholes.

A theory of rational choice that appeals to preferences and expectations—desires and beliefs—could not be even a model with the power of the simplest Mendelian variety.[6] Such a theory is intentional psychology formalized, and so suffers from all the problems that bedevil its appeal to intentional content. This means that economics, political science, and those parts of sociology and psychology that exploit the framework of game theory are fated to provide case studies, models of far greater idealization than anything to be met with in population biology, generalizations of very limited scope and predictive power, and a great deal of human history.[7] Perhaps social science may hope

---

5. For a clear example of this sort of condemnation of the possibility of behavioral biology, see John Horgan, "Trends in Behavioral Genetics," *Scientific American*, July 1993, pp. 122–131.

6. For a detailed discussion, see Alexander Rosenberg, *Economics* (Chicago: University of Chicago Press, 1992), chapters 4 and 5.

7. It is ironic that these rational-choice assumptions and their game-theoretical instantiations have had interesting results in evolutionary behavioral biology, where they have shed light on the evolution of reciprocal altruism and other strategies of cooperation and competition. See John Maynard Smith, *Evolution and the Theory of Games* (Cambridge: Cambridge University Press, 1982).

for some sketch of a part of the underlying mechanism for large-scale social processes, especially those in which individual agents have at least some incentive to adapt their behavior to a prudential interpretation of principles of economic rationality.

It is by no means clear that the social sciences could do better in the provision of generalizations and theories were they to eschew intentionality in the description and explanation of human behavior. Though they might provide more robust models and more reliable generalizations, the subject matter of these generalizations and models would no longer be of interest to those who look to social science for useful information. Given our cognitive and computational limits, and our interests as biological and social creatures, biology turns out to be a very useful tool, so useful that we are inclined to mistake it for a science true about the world and independent of us. The same cannot be said for the social sciences. They certainly are not true independent of us and our existence, and they cannot avoid being less useful instruments than biology for agents like us.

Bibliography

Beatty, John. "Optimum Design Models and the Strategy of Model Building in Evolutionary Biology," *Philosophy of Science* 47 (1980): 532–561.

———. "Chance and Natural Selection." *Philosophy of Science* 51 (1984): 183–211.

———. "Insights and Oversights of Molecular Biology." In Michael Ruse, ed., *The Philosophy of Biology.* New York: Macmillan, 1989.

Beatty, John, and Finsen, Susan. "Rethinking the Propensity Interpretation: A Peek inside Pandora's Box." In Michael Ruse, ed., *What the Philosophy of Biology Is,* pp. 17–30. Boston: Kluwer, 1987.

Bennett, Jonathan. *Linguistic Behavior.* Cambridge: Cambridge University Press, 1976. Reprint, Indianapolis: Hackett, 1989.

Brandon, Robert. *Adaptation and Selection.* Princeton: Princeton University Press, 1990.

Churchland, Paul. *Matter and Mind.* Cambridge: MIT Press, 1984.

Cooper, W. S. "Expected Time to Extinction and the Concept of Fundamental Fitness." *Journal of Theoretical Biology* 107 (1984): 603–629.

Darwin, Charles. *On the Origin of Species.* London: John Murray, 1859.

Davidson, Donald. *Essays on Action and Events.* Oxford: Oxford University Press, 1981.

Dawkins, Richard. *The Selfish Gene.* New York: Oxford University Press, 1976.

Dennett, Daniel. *Content and Consciousness.* London: Routledge and Kegan Paul, 1969.

———. *Brainstorms.* Montgomery, VT: Bradford Books, 1978.

———. "Intentional Systems in Cognitive Ethology: The Panglossian Paradigm Defended." *Behavioral and Brain Sciences* 6 (1983): 343–390. Reprinted in *The Intentional Stance.* Cambridge: MIT Press, 1987.

———. *The Intentional Stance.* Cambridge: MIT Press, 1987.

———. "Real Patterns." *Journal of Philosophy* 88 (1991): 27–51.

Dretske, Fred. *Knowledge and the Flow of Information.* Cambridge: Bradford Books, MIT Press, 1981.

———. *Explaining Behavior: Reasons in a World of Causes.* Cambridge: Bradford Books, MIT Press, 1988.

Dupré, John. *The Disorder of Things: Metaphysical Foundations of the Disunity of Science.* Cambridge: Harvard University Press, 1993.

Feynman, Richard. *QED: The Strange Theory of Light and Matter.* Princeton: Princeton University Press, 1986.

Fodor, Jerry. *A Theory of Content and Other Essays.* Cambridge: MIT Press, 1990.

Gaspar, Philip. "Reductionism and Instrumentalism in Genetics." *Philosophy of Science* 59 (1992): 655–670.

Hempel, Carl. *Aspects of Scientific Explanation and Other Essays.* New York: Free Press, 1965.

Horgan, John. "Trends in Behavioral Genetics: Eugenics Revisited." *Scientific American,* July 1993, pp. 122–131.

Horgan, Terry, and Woodward, James. "Folk Psychology Will Always Be with Us." *Philosophical Review* 94 (1985): 197–226.

Hull, David L. *The Philosophy of Biological Science.* Englewood Cliffs, NJ: Prentice Hall, 1974.

————. *Science as a Process.* Chicago: University of Chicago Press, 1989.

Jammer, Max. *Concepts of Mass in Classical and Modern Physics.* Cambridge: Harvard University Press, 1961.

Kincaid, Harold. "Molecular Biology and the Unity of Science." Manuscript. 1991.

Kitcher, Patricia. "In Defence of Intentional Psychology." *Journal of Philosophy* 81 (1984): 89–106.

Kitcher, Philip. "1953 and All That: A Tale of Two Sciences." *Philosophical Review* 93 (1984): 335–373.

————. "Explanatory Unification and the Causal Structure of the World." In Philip Kitcher and Wesley Salmon, eds., *Scientific Explanation: Minnesota Studies in the Philosophy of Science,* 13: 410–505. Minneapolis: University of Minnesota Press, 1989.

Lawrence, Peter A. *The Making of a Fly.* Cambridge, MA: Blackwell Scientific, 1992.

Lewis, David. *Philosophical Papers. Volume 2.* Oxford: Oxford University Press, 1986.

Lewontin, Richard. *The Genetic Basis of Evolutionary Change.* New York: Columbia University Press, 1974.

Lewontin, Richard, and Gould, Stephen J. "The Spandrels of San Marco and the Panglossian Paradigm: A Critique of the Adaptationalist Programme." *Proceedings of the Royal Society of London* 205 (1979): 581–598.

Li, Wen-hsiung, and Graur, Dan. *Fundamentals of Molecular Evolution.* Sunderland, MA: Sinauer, 1991.

Lloyd, Daniel. *Simple Minds.* Cambridge: Bradford Books, MIT Press, 1989.

Lloyd, Elisabeth. *The Structure and Confirmation of Evolutionary Theory.* New York: Greenwood, 1989. Reprinted, Princeton: Princeton University Press, 1993.

McCloskey, Donald. *The Rhetoric of Economics.* Madison: University of Wisconsin Press, 1985.

Maynard Smith, John. *Evolution and the theory of Games.* Cambridge: Cambridge University Press, 1982.

Mill, John Stuart. *System of Logic.* 1849. In *Collected Works.* Toronto: University of Toronto Press, 1896.

Millikan, Ruth. *Language, Thought, and Other Biological Categories: New Foundations for Realism.* Cambridge: MIT Press, 1984.

Mills, Susan, and Beatty, John. "The Propensity Definition of Fitness." *Philosophy of Science* 46 (1979): 263–286.

Monod, Jacques. *Chance and Necessity.* New York: Knopf, 1971.

Nagel, Ernest. *The Structure of Science.* New York: Harcourt, Brace, World, 1961. Reprint Indianapolis: Hackett, 1983.

Putnam, Hilary. *Mind, Language, and Reality.* Cambridge: Cambridge University Press, 1975.

Quine, Willard V. O. *From a Logical Point of View.* Cambridge: Harvard University Press, 1953.

———. *Word and Object.* Cambridge: MIT Press, 1960.

Robinson, Joseph. "Reduction, Explanation, and the Quest for Biological Knowledge." *Philosophy of Science* 53 (1986): 333–353.

Rosenberg, Alexander. "The Supervenience of Biological Concepts." *Philosophy of Science* 45 (1978): 368–386.

———. *The Structure of Biological Science.* Cambridge: Cambridge University Press, 1985.

———. *Economics: Mathematical Politics or Science of Diminishing Returns?* Chicago: University of Chicago Press, 1992.

Ruse, Michael. *The Philosophy of Biology.* London: Hutchinson University Library, 1973.

Schaffner, Kenneth. "Approaches to Reduction." *Philosophy of Science* 34 (1967): 137–147.

Sober, Elliott. "Frequency-dependent Causation." *Journal of Philosophy* 79 (1982): 247–253.

———, ed. *Conceptual Issues in Evolutionary Biology.* Cambridge: MIT Press, 1984.

———. "Fact, Fiction, and Fitness." *Journal of Philosophy* 81 (1984): 372–382.

———. *The Nature of Selection.* Cambridge: MIT Press, 1984. Reprinted, Chicago: University of Chicago Press, 1993.

Sober, Elliott, and Lewontin, Richard. "Artifact, Cause, and Genic Selection." *Philosophy of Science* 49 (1982): 157–180.

Sterelny, Kim, and Kitcher, Philip. "The Return of the Gene." *Journal of Philosophy* 85 (1988): 339–362.

Sterelny, Kim; Kitcher, Philip; and Waters, C. Kenneth. "The Illusory Riches of Sober's Monism." *Journal of Philosophy* 97 (1990): 158–161.

Strickberger, Monroe. *Genetics.* First edition. New York: Macmillan, 1968.

Thoday, J. M. *Symposia for Society for Experimental Biology.* 7 (1953): 96–113.

Thompson, Paul. *The Structure of Biological Theories.* Albany: SUNY Press, 1988.

Toulmin, Stephen. *The Philosophy of Science*. London: Hutchinson University Library, 1953.

Van Fraassen, Bas. *The Scientific Image*. Oxford: Oxford University Press, 1979.

Waters, C. Kenneth. "Natural Selection without Survival of the Fittest." *Biology and Philosophy* 1 (1983): 207–225.

———. "Environment, Pragmatics, and Genic Selectionism." *Proceedings and Addresses of the American Philosophical Association* 59 (1985): 359.

———. "Why the Anti-reductionist Consensus Won't Survive: The Case of Classical Mendelian Genetics." In *PSA 1990*, pp. 125–139. East Lansing, MI: Philosophy of Science Association, 1990.

———. "Tempered Realism about the Forces of Selection." *Philosophy of Science* 58 (1991): 553–573.

———. "Genes Made Molecular." *Philosophy of Science*, forthcoming.

Williams, G. C. *Adaptation and Natural Selection*. Princeton: Princeton University Press, 1966.

———. *Evolution: Selected Papers*. Ed. William B. Provine. Chicago: University of Chicago Press, 1984.

Williams, Mary B. "Deducing the Consequences of Evolution." *Journal of Theoretical Biology* 29 (1970): 343–385.

Index